Sustainability in the Built Environment

"This book pursues a question that ought to matter to everyone: how can buildings contribute to comfortable human life on our planet? To answer theis question the book examines a wide range of topics related to sustainable buildings, in language accessible for a general audience. And it features some valuable content not easily found elsewhere. I hope property professionals will read this book and implement some of its good ideas within their organizations!"

—John Haffner, *Deputy Director—Sustainability, Hang Lung Properties, Co-Chair, APAC Net Zero Council, Urban Land Institute, and Global Governing Trustee, Urban Land Institute*

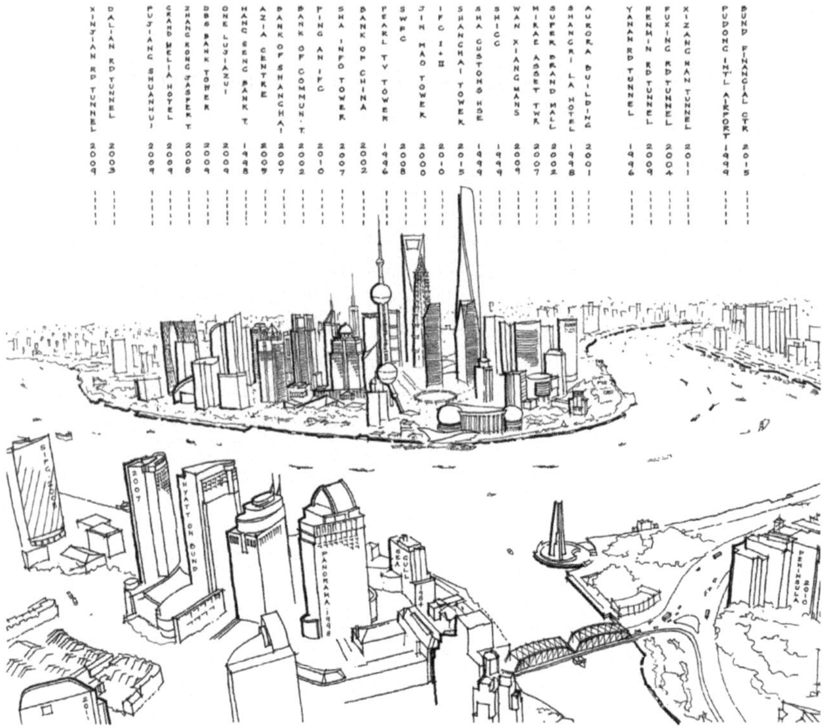

8/2014 No. /35 · Edition II 'In China Since 1994.' Stephen White

Sam Crispin

Sustainability in the Built Environment

The Case for Green Buildings in Rapidly Urbanising Economies

palgrave
macmillan

Sam Crispin
Hong Kong, China

ISBN 978-981-95-6548-1 ISBN 978-981-95-6549-8 (eBook)
https://doi.org/10.1007/978-981-95-6549-8

This Palgrave Macmillan imprint is published by the registered company Springer Nature Singapore Pte Ltd.
The registered company address is: 152 Beach Road, #21-01/04 Gateway East, Singapore 189721, Singapore

Thank you to my parents Jane and Julian and my family, Maisie, Chloe and Felix. Your love, support and understanding knows no bounds.

Gratitude also to all those professionals past, present and future, who have, are and will work tirelessly towards sustainability in the built environment and greener buildings everywhere.

Acknowledgements This book would not have been written without support from family. Thank you all.

Special thanks to Dr. Paul Clark of Thompson Rivers University for his never ending support and encouragement. Thanks also to Mark O'Neil and Hugh Peyman for editorial assistance and support.

Frontispiece image courtesy of and © Stephen White for which very many thanks.

Contributors

Thanks to all the interviewees and contributors who were kind enough to spare their valuable time time and share their even more valuable knowledge and support. Sometimes a response to a question or request for a quote validates all the late nights.

Andy To
Cary Chan
Dhawal Doshi
Fred Pilbrow
Hannah Jeong
Ian Booth
Ian Riley
Iris Kampers
James Brearley
John Haffner
Jonathan Hannam
Professor Khee Poh Lam
Professor Lapman Lee
Lizette McNeill
Mark Bawtree
May Lv
Melanie Alshab
Peter Dampier
Raefer Wallis
Robert Hansor
Stephane Hallegatte

Stephen White
Sunny Zhang
Vinita Turakhia
Professor Xiaoling Zhang
Not forgetting those who supported but chose to remain anonymous and the authors of materials I quoted from published sources.

Competing Interests The author has no competing interests to declare that are relevant to the content of this manuscript.

Introducing the Author

A native of the United Kingdom, I have spent more than 30 years working in the property (real estate) industry, and lived in Shanghai, Hong Kong and Taipei for 31 of the last 35 years. When I moved from Shanghai to Hong Kong in 2017 I had spent half my life living in Shanghai. In 2021, I obtained a Masters' degree in Corporate Environmental Governance at The University of Hong Kong. From 2022 to 2025 I was Senior Director, Regional Sustainability & Environment, Social and Governance (ESG) at a property consultancy, based in Hong Kong. In this role I managed both internal Asia Pacific ESG commitments and developed a range of sustainability services for existing and new clients.

My interest in sustainability started with surveying the flood risk and mitigation efforts of the River Ravensbourne in Catford, southeast London as part of a Geography lesson at the age of 14 or 15 (thank you Mr Irvine). My interest in Geography was initially based on getting out of the classroom to do something more interesting, but soon I found deeper interest. I went on to survey longshore drift on Hayling Beach as a later A Level project, improvising research methodology as I went. Later, my undergraduate dissertation was on the "Environmental and Social Costs and Benefits of the 3 Gorges Dam in China", the world's biggest hydropower project. This topic combined my study of both Chinese and Geography. I concluded that the project would proceed regardless of the E or the S because of the G. This conclusion came to reality as my work was being marked in 1992. Only many years later did I realise that this effort was an early version of what we might now describe as an ESG report. It was written with an unsuitable framework borrowed from the world of business.

After completing my master's degree in 2021 I found to my astonishment the extent of built environment related emissions and felt the need to document how lasting and meaningful change could be accomplished. It is mainly intended as a guide to the uninitiated. Industry leaders, professional consultants, property executives, letting agents, business executives from other industries and students of both real estate and sustainability may all find something of value in these pages.

I wanted to put on record the significant environmental issues of the buildings and infrastructure that we all depend on and ask the critical question, "why are so few people talking about sustainability in the built environment?" At the same time I wanted to demonstrate the solid business, public health and environmental reasons for everyone to do something and indicate a path to overcome the various obstacles to taking meaningful action. This book is intended to help get the discussion going so we can talk more about it. We all have a role to play.

This is not a technical book, I am not a building engineer, and it does not provide a "how to" guide to making buildings more sustainable. I'd like to think it is bigger than that and offers detailed strategic insight into how an entire industry supply chain can be transformed at scale using proven models for implementation, particularly in rapidly urbanising economies, while improving returns on investment. The megacities of the 21st century will be built in Asia and Africa in the coming decades. It is here that we must apply our lessons, implement the necessary changes and build sustainably. For this reason the content meanders through a wide range of subject matter offering glimpses into rabbit holes. I have always tried to pull it back to weave around the core subject which is the business and environmental case for green buildings. The data suggests that this is indeed something compelling, urgent and that we should all be asking more questions about. Demand for change starts with asking difficult questions.

While the basic principles are straightforward, some of the subject matter is complex and full of acronyms. I have chosen not to go into great detail on the complexities but have touched on them for the reader to follow up on according to their interest. Not all the subject matter fits neatly into a single chapter and for this reason there is overlap between different sections which some readers may find confusing, but I have had to take a reasonable view on this.

At the same time, effort has been made to avoid unnecessary jargon throughout but some is unavoidable. The following notes may help readers who do not have a background in real estate, buildings and sustainability. In most cases I have provided further supporting explanations in the text.

1. CO2, carbon dioxide, is a gas that occurs naturally.
2. CO2 emissions is a term that describes the release of CO2 gas as a result of human activities.
3. CO2e means carbon dioxide equivalent and is used to measure different greenhouse gases in a measure that is equivalent to CO2.
4. Greenhouse Gases are gases (including water vapour) that trap solar energy within the earth's atmosphere. Some of these gases occur naturally and some are manmade. We need some greenhouse gases but not too many.
5. In late August 2025, the global average CO2 in the atmosphere had reached over 420 parts per million. This compares to a pre-industrial average of around 280 ppm.
6. Almost all scientists agree that anthropogenic climate change is a scientific fact, like all science it can be and is disputed.
7. Different data sources may provide slightly different figures for the same or similar things, e.g. total energy related carbon emissions from the built environment which varies from year to year.
8. The ratio of embodied and operational carbon in buildings depends on each individual building. Location will affect the material and energy inputs, or there may be simply different measurement boundaries or counting methodology. The data generally shows a consistent result within a few percentage points, if not there is usually a rational explanation.
9. The term "built environment" is defined at the start of Chapter 3 but I have used the term fairly interchangeably with "buildings and construction" throughout this book to refer to all buildings and manmade structures. The focus of the content is largely building related, specifically commercial buildings. Most of the principles discussed can be applied to infrastructure, residential buildings and so on.

10. I have used square meters (sq.m) as a standard building measurement and all currency has been equalised to US$. Both building measurement standards and exchange rates vary.
11. There has been no primary research conducted in preparing this book. The general direction was set before the research was completed based my knowledge and experience of the subject. Actual contents have been revised along the way according to the subject matter. This has resulted in a more optimistic book than I had first envisaged.
12. I have taken care to find what I consider reliable and trustworthy sources but not all are peer reviewed academic papers. In some cases I have considered alternative sources before inclusion but only one source is mentioned in the references.
13. Sources include businesses involved in promoting products and services, government and government related websites, magazines and newspapers. No single source has been relied on excessively.
14. Interviewees are mostly people known personally to me with the relevant expertise, this may result in a degree of bias. Some attributed comments are taken from publicly available sources and I have indicated as such, a couple of sources did not wish to be named and we must respect that.

Contents

Abbreviations

ACEE	American Council for an Energy-Efficient Economy
AIA	American Institute of Architects
APAC	Asia Pacific
BAS	Building Automation System
BCA	Building Construction Authority
BEAM	Building Environment Assessment Method
BEC	Building Envelope Campaign
BESS	Battery Energy Storage System
BIM	Building Information Modelling
BIMCO	Baltic and International Maritime Council
BIPV	Building Integrated Photovoltaics
BMS	Building Management System
BMU	Building Maintenance Unit
BOMA	Building Owners and Managers Association
BRE	Building Research Establishment
BREEAM	Building Research Establishment Environmental Assessment
BRI	Building Resilience Index
CABE	Chartered Institute of Building Engineers
CAGR	Cumulative Average Growth Rate
CAPEX	Capital Expenditure
CBD	Central Business District
CCUS	Carbon Capture Utilisation and Storage
CDO	Chief Development Officer
CDP	Carbon Disclosure Project
CE100	Circular Economy 100
CFC	Chlorofluorocarbon

CFE	Carbon Free Energy
CIBSE	Chartered Institute of Building Service Engineers
CLP	China Light and Power
CLT	Cross Laminated Timber
CO2	Carbon Dioxide
COO	Chief Operating Officer
CORSIA	Carbon Offsetting and Reduction Scheme for International Aviation
CPCF	Construction Productivity and Capability Fund
CPD	Continuing Professional Development
CRREM	Carbon Risk in Real Estate Monitor
DCF	Discounted Cash Flow
DfMA	Design for Manufacture and Assembly
DoE	Department of Energy
EAF	Electric Arc Furnace
ECBC	Energy Conservation Building Code
ECM	Energy Conservation Measure
EEXI	Energy Efficiency Existing Ship Index
EIA	Environmental Investigation Agency
EIT	European Institute of Innovation and Technology
EMF	Ellen MacArthur Foundation
EPC	Energy Performance Certificate
EPRD	Energy Performance of Buildings Directive
EROEI	Energy Return on Energy Investment
ESG	Environment, Social and Governance
ESRT	Empire State Realty Trust
EU	European Union
EV	Electric Vehicle
FDRA	Footwear Distributors and Retailers of America
FSC	Forest Stewardship Council
GCC	Gulf Cooperation Council
GCCA	Global Cement and Concrete Association
GDP	Gross Domestic Product
Gha	Global Hectares
GGBFS	Ground Granulated Blast Furnace Slag
GHG	Greenhouse Gas
GRESB	Global Real Estate Sustainability Benchmark
GTS	Growth and Transformation Scheme
GWh	Gigawatt Hours
GWP	Global Warming Potential
HCFC	Hydrochlorofluorocarbon
HFC	Hydrofluorocarbon
HKGBC	Hong Kong Green Building Council

HVAC	Heating, Ventilation and Air Conditioning
ICA	Insurance Council of Australia
ICC	International Code Council
IEA	International Energy Agency
IEC	Illuminating Engineering Council
IEQ	Indoor Environmental Quality
IFC	International Finance Corporation
IFMA	International Facility Management Association
IGCC	International Green Construction Code
IGSD	Institute for Governance and Sustainable Development
IMO	International Maritime Organisation
IPCC	Intergovernmental Panel on Climate Change
ISEP	Institute of Sustainability and Environmental Professionals
IVS	International Valuation Standards
IXO	Inside to Outside
JLL	Jones Lang LaSalle
KL	Kuala Lumpur
kWh	Kilowatt-hour
LCA	Life Cycle Assessment
LED	Light Emitting Diode
LEED	Leadership in Energy and Environmental Design
LRM	Lifecycle Refrigerant Management
LTAG	Long Term Aspirational Goal
LVMH	Moët Hennessy Louis Vuitton
M&S	Marks and Spencer
MEP	Mechanical, Electrical and Plumbing
MIT	Massachusetts Institute of Technology
MOQ	Minimum Order Quantity
MoU	Memorandum of Understanding
NA	Natural Aggregates
NABERS	National Australian Built Environment Rating System
NAREIT	National Association of Real Estate Investment Trusts
NCREIF	National Council of Real Estate Investment Fiduciaries
NDC	Nationally Determined Contribution
NEA	National Environment Agency
NEECA	National Energy Efficiency and Conservation Authority
NGO	Non-Governmental Organisation
NRDC	National Resources Defense Council
NZE	Net Zero Emissions
NZERB	Net Zero Emission and Resilient Buildings
OPEX	Operating Expenses
PEEB	Partnership for Energy Efficiency in Buildings
PRI	Principles for Responsible Investment

RA	Recycled Aggregates
RAC	Refrigeration and Airconditioning
RAS	Recirculating Aquaculture System
REC	Renewable Energy Certificate
RIBA	Royal Institute of British Architects
RICS	Royal Institution of Chartered Surveyors
RMI	Rocky Mountain Institute
SAF	Sustainable Aviation Fuel
SARS	Severe Acute Respiratory Syndrome
SCB	Standard Chartered Bank
SDG	Sustainable Development Goal
SGP	Singapore Green Plan
SGBMP	Singapore Building Masterplan
SLE	Super Low Energy
SOS	Sustainable Office Solutions
SPiRiT	Sustainable Project Rating Tool
Sq.m	Square metre
SWNZ	Sustainable Winegrowing New Zealand
TBL	Triple Bottom Line
TWh	TerraWatt-hour
UCEM	University College of Estate Management
UCL	University College London
ULI	Urban Land Institute
UNDP	United Nations Development Programme
UNEP	United Nations Environment Programme
UNESCO	United Nations Educational Scientific and Cultural Organisation
USACE	United States Army Corps of Engineers
VSD	Variable Speed Drive
VW	Volkswagen
WCA	World Cement Association
WCED	World Commission on Environment and Development
WEF	World Economic Forum
WGBC	World Green Building Council
WHO	World Health Organisation
WINT	Water Intelligence
WMBC	We Mean Business Coalition
WRI	World Resources Institute

List of Figures

List of Tables

CHAPTER 1

Preface—Humanity's Future Depends on the Building and Construction Industry

"We never know the worth of water till the well is dry" Thomas Fuller, seventeenth century.

Our lives on this planet depend on the building and construction industry, including roads and infrastructure. At the same time, the building and construction industry is the most ruinous for the environment. Building materials, construction, operation and use and demolition of buildings vary by year but consistently account for between 34 and 38% of CO_2 emissions in recent years. The industry sucks in larger quantities of material, labour and energy resources than any other. Real estate development and use directly and indirectly contribute more man-made Greenhouse Gases (GHG) than any other human activity. But we cannot live without cities. According to the World Bank, they account for about 80% of global GDP, more than 70% of global CO_2 emissions and about 50% of global waste while occupying only 3% of the world's land area. The United Nations predicts that, by 2050, about 68% of the global population will be urban residents. The World Economic Forum (WEF) estimates the climate crisis may result in 14.5 million deaths and US$12.5 trillion in economic losses by 2050. Reducing built environment-related emissions by, say, 50% can make a dramatic impact on global emissions now and in future as future urbanisation takes place. So why do we hear

S. Crispin, *Sustainability in the Built Environment*,
https://doi.org/10.1007/978-981-95-6549-8_1

so little about the environmental impact of the built environment that sustains us?

If we cannot transition existing cities and build new ones to be environmentally sustainable, greenhouse gas emissions will increase dramatically, temperatures will rise, climate disasters will increase and, finally, the planet will no longer be habitable for as many humans as it is today. That is the subject of this book. Do we have the means and the knowledge to construct buildings and cities that are sustainable? Do we have the political and financial will? The answer to the first question is "Yes", while the answer to the second remains uncertain. A key third question remains: why is sustainability in the built environment barely talked about compared to, say, electric vehicles, food production, aviation or ocean plastic? This book addresses these questions and explains what we must do.

Cities are essential for economic growth, efficient provision of housing, health and education services to support large and growing populations. We must create buildings for people to live and work in. Unfortunately, including even wood, all the materials used in modern buildings require energy intensive extraction, processing and manufacture. They include steel, concrete and concrete blocks, aluminium, glass, wood, plasterboard and insulation. Other inputs include water, labour and other forms of energy. They are also financed and insured. Short of ceasing all development and radically changing production processes, there is no way to change the demand for, and the nature of, these inputs for the built environment.

What we can change is the way we create cities and buildings by following best practices. This includes using materials that, at the end of their life cycle, can be recycled and reused; creating buildings that use renewable energy and with design, layout, orientation and structure that are energy efficient; and creating urban areas that have a high area of tree cover that reduces temperatures and extreme heat-related deaths. Various case studies from a variety of geographies are used in different chapters to briefly show what can be and has been achieved in different projects.

While technical topics are covered, this book is not a technical guide; those details are available elsewhere. The book does not either explore some of the more radical theories such as degrowth, which bases sustainability solutions around reduced consumption. Degrowth may be a laudable objective, but consumer societies are still moving in the opposite direction. Some readers may find these gaps to be a limitation, but the breadth of topics is too wide to go into full detail on everything.

What is lacking in technical detail may be compensated for by the interdisciplinary approach encompassing business and finance, environmental science, sociology and public policy. This means that not all sections fit neatly into a siloed chapter; some sections overlap with those found in earlier chapters.

The overarching aim was to use accessible language to describe a strategic approach to the holistic transition that is needed for the built environment as an industry, including the extensive supply chain, to drastically reduce emissions. It explains in detail the steps governments, institutions, companies and individuals need to take to improve and reform the building industry and create the "virtuous cycle" that is sustainable and will go a long way to preserving comfortable human life on this planet. At various sections, the content reverts to the business case, and examples of what has already been achieved in different projects from a variety of locations are embedded into the relevant sections to support the business case.

CHAPTER 2

The Business Case—Property Values

"Managers and investors alike must understand that accounting numbers are the beginning, not the end, of business valuation" Warren Buffet.

ESG and Valuation Guidelines

The transformation of the built environment from the "business as usual" carbon intensive model to a low carbon model takes investment. As with all investment, a return is expected. The existence or prospect of carbon pricing as well as future proofing against current and future climate-related risks are all factors to be included in financial modelling and valuations.

From January 2025, the International Valuation Standards (IVS) has mandated the inclusion of ESG-related factors in valuations "the impact of significant ESG factors should be considered in determining the value of a company, asset or liability", this includes qualitative and quantitative impacts, risks and opportunities but need to be measurable and reasonable, according to the valuers professional judgement (Aronsohn, 2024). The Royal Institution of Chartered Surveyors (RICS) is a body of professional members that "help to create and protect built and natural environments that are sustainable, resilient and inclusive for all" (RICS, 2024). A key document in the incorporation of ESG data into property

S. Crispin, *Sustainability in the Built Environment*,
https://doi.org/10.1007/978-981-95-6549-8_2

valuations was published by the RICS in February 2024. Described as "a practical reference document on legislative, market-driven and future ESG requirements for valuers and financial clients in the EU", it seeks to address the challenges the built environment faces in moving from "believe" and "suggest" to evidence and data-based facts where ESG is explicitly considered in valuations (Scherrenberg et al., 2024).

An important consideration in determining the ESG impact on value is the valuation methodology. These include the comparison or market approach using available market comparables and adjusting the data based on the actual versus the comparable. The income approach takes the income from all revenue streams and capitalises it to provide a value. The cost approach is based on the replacement cost, including land, construction, profit and all other costs of an identical building. According to the RICS document, ESG indicators may have an impact on cash flow, including higher rents, lower operating costs, green finance, etc., which are cash flow drivers. Or they may be risk drivers that reduce risk from impact of future regulations, improving resilience to physical risks such as flooding and so on, which may impact the market value. Finally, value drivers, which include green certification, sustainable design, low emissions and so on, result in higher rents and market values.

This is not all. Additional future ESG indicators mentioned include waste management, carbon footprint, biodiversity, indoor air quality, affordability, community impact, DEI, water conservation, lighting and safety. In reality, many of these may have been included in Table 2.1 in some way. There is also a risk of double counting, for example, renewable energy generation and greenhouse gas emissions. This needs to be managed when allowing for these factors.

Iris Kampers is Director, ESG and Sustainability Savills Netherlands Holding BV. She points out that the RICS ESG factors are interlinked, such as Greenhouse Gas Emissions and on-site energy generation, and comments that while energy label A building valuations have gone up, B and C-rated building valuations have gone down (the EU Energy Performance Certificate scale is from A to G). According to Kampers, tenants are increasingly stating expectations for an A while suggesting there are limited returns on investment in the next step, investing to gain a net zero rating.

Table 2.1 RICS "ESG indicators"

Indicator	*Data and units*	*Remarks*	*Impact on value*
Energy rating & Energy consumption	Energy use intensity kWh/sq.m Local market ratings, e.g. EPC, NABERS Energy audits	Regulatory compliance and anticipated regulatory changes Covers common and leased areas	Improved cash flow and reduced risk Benchmark with comparable buildings
On-site renewable energy generation	Type in kWh/sq.m % of total consumption Heating	Meet local market expectations Attractive to tenants	Improves operating efficiency Improves cash flow
Green building certifications	Includes LEED, BEAM + , Green Star, NABERS, WELL, etc	Increasingly a basic tenant requirement for building selection Exact certifications vary in nature including design vs operations	Based on a range of factors that attract tenants. Can reduce risk of void periods Reduces risk and may improve value depending on exact nature of certification and local market
Greenhouse Gas Emissions	Electricity use Fossil fuel use Refrigerants and other chemicals Kg CO2 equivalent per sq.m per year	Electrification supports removal of all fossil fuel from the building operating supply chain Work with certified contractors for sustainable cleaning, refrigerant disposal etc	A key benchmarking data point to compare with others Lower GHG emissions helps future proof the building. Supports higher comparable valuation
Decarbonisation pathway	Energy reduction plans Net zero targets Retrofitting budgets	Costed and time bound implementation plans	Operating efficiency improvements Lower risk and higher comparable valuation
Physical risk	Climate risk analysis and mitigation plan	Budget for mitigation measures	Reduced insurance costs Lower risk and higher comparable valuation

(continued)

Table 2.1 (continued)

Indicator	*Data and units*	*Remarks*	*Impact on value*
Location	Transport connectivity especially mass transit and other public transport Supporting facilities for occupants including those in nearby buildings	Improves demand for the building by making it more acceptable to a wider range of employees Considered at building selection stage	Makes property more attractive and increases rents Reduces void periods
Sustainable transport	Active travel facilities, e.g. bike racks per employee EV charging points per employee	Use of low carbon transport supports occupier scope 3 targets	A consideration for tenants mindful of staff travel options and reducing scope 3 emissions
Access for all	Step free access	Retrofitting of older buildings may be required	A requirement for some tenants
Green lease agreements	Activates sustainability collaboration between landlord and tenant % of tenants participating Building ESG Committee	Tailor made to meet stakeholder needs Smaller tenants can benefit from efficiency gains by participating in building wide activities Meet tenant expectations to drive sustainability performance	Increase tenant satisfaction reducing void periods Increase cash flow and capital values
Embodied carbon	Sustainable design Documented use of low carbon materials Circularity metrics, e.g. % recycled content, landfill diversion rates	Sustainable fit out guide applies to all tenants Certification of tenant fit outs and landlord upgrades Improve building competitiveness, depending on local market	Part of a comparable valuation approach

(Scherrenberg et al., 2024).

Valuation guidelines are one thing, what about the market evidence for higher rents in green buildings? This is difficult to determine with any certainty considering that each building is unique and there are a large number of different factors that influence each deal. According to Hannah Jeong, Executive Director and Head of Valuation & Advisory Services at CBRE Hong Kong, there is often insufficient market evidence to support premium values using the market or comparable valuation approach. There may be a long list of sustainable attributes to a building but unless there is evidence to show premiums or cost savings, any relationship between the shade of green of the building and the valuation may be entirely coincidental. At the same time, landlords that care about sustainability probably also demonstrate management excellence and pay more attention to their tenants.

In Switzerland, discounted cash flow (DCF) is the main valuation method for investment property. When it comes to putting a discounted cash flow together, the following data points are explicitly affected by sustainability considerations:

1. Shorter void periods between tenants.
2. Higher rents.
3. Lower risk of vacancy.
4. Improved renewal terms.

In a DCF valuation, ESG factors can be incorporated using added or integrated approaches. Quantifying the impact on cash flow of the sustainability considerations, such as additional revenues from renewable energy generation, enables them to be added into the DCF in a transparent manner and isolated if necessary. Assumptions around rent increases and void periods can be beneficial for more sustainable buildings (Ganz, 2024).

Green Premiums

The growing evidence is that there are green premiums that reflect improved sustainability performance and brown discounts for those that don't. The shade of green is also important. The CBRE UK Sustainability Index to the end of 2024 shows that in the office sector, efficient buildings outperformed inefficient ones when measured by total returns, capital growth and rental value growth. The same report was unable to make the same clear distinction for retail, and in the industrial sector, the inefficient buildings showed similar rental value growth as inefficient ones (CBRE, 2024).

Finding individual buildings where those with green certifications have higher rents is not difficult. A Savills report in 2022 identified a rental premium of around 18% in Bangkok and Singapore but only 4 and 2% in Kuala Lumpur and Tokyo respectively (Savills, 2022). While Hannah Jeong's view is that in Hong Kong's oversupplied office market green premiums are around 3%.

The evidence from the market research reports is that highly certified assets not only get higher rents, they also have lower void periods between tenants, enjoy lower operating costs, benefit from green finance and secure longer term capital values.

Attributing a premium rent directly to a green certification, or particular green features, is much more nuanced; some of the figures in Table 2.2 have been adjusted to account for some of the variables. The biases at play include that new buildings tend to attract rent premiums and they also tend to have higher green certifications. Also it is hard to know how much of the premium is due to being green, being new or other factors such as location, access to transit and non-rent contractual terms such as incentives for new tenants.

Academic research also attempts to account for these biases. Onishi et al., analysed 37,346 green-certified buildings in Tokyo, "a large heterogeneous market with advanced environmental technology". They found a 6.5% premium for green-certified buildings but acknowledge that the biases mean that this is an overestimate. Sub samples reveal a 5.4% green premium for medium-sized old buildings, which, as existing buildings, will have required retrofitting to achieve green certifications, and 2.6% premium for large-sized new buildings, which will have been green from the ground up. The study also notes that the green premiums will decrease over time as green certifications become the norm(Onishi et al., 2021). Although the focus may change to shades of green or specific metrics, such as energy use intensity, the diminishing financial incentive to be green will be a problem for regulators to tackle.

The Global Real Estate Sustainability Benchmark (GRESB) is a platform to assess real estate portfolio sustainability performance. According to a GRESB report "Key financial benefits of GRESB participation",

Table 2.2 Summary of reported green premiums

Organisation	*Green Premium*	*Market*	*Comment*
CBRE	3%	Asia	Small number of occupiers willing to pay slightly higher rents, sizable number willing to pay marginal premium
CBRE	3.7–5.5%	US and EU (meeting energy regulations is a factor in nearly 75% of cases, carbon neutrality a factor in 54%)	Globally, around 40% of occupiers would pay a premium for superior resiliency, over 20% would expect a discount if no superior resiliency
JLL	Up to 22% in markets like Hong Kong, 7% in Singapore	Asia	Based on evaluation of 6493 buildings in 41 cities in Asia Pacific
Knight Frank	Up to 17.1%	Hong Kong	Based on 181 buildings in Hong Kong and adjusted for age and location
Cushman & Wakefield	10–30% higher for logistics properties	Europe	19–24% price premium Based on analysis of 1500 transactions over 4 years from 2019
WEF	Rent premium of 6% Sales premium of 7.6%	Global (2022)	Based on analysis of 42 studies by Dalton and Fuerst (2018)
UBS	28% New York 19% London	New York and London	Analysis of 1453 building transactions, adjusted for various factors Recommend 1–3% brown discount to annualised return allowance for retrofit costs.

(CBRE, 2023a, 2023b; JLL, 2022; Wong et al., 2024; Coleman, 2024; Ulbrich, 2022; Chen, 2023)

sustainability practices can improve competitiveness, contribute to financial performance and attract investors. They list the following benefits:

1. Potential reduced borrowing costs through sustainability-linked loans.

2. Attracting investors. 44% of investors mandate GRESB participation in the investment contracts, with some setting specific targets.
3. Sustainability performance in the form of energy efficiency and resource management leads to operational efficiency.
4. Contribute to long-term financial stability by supporting ESG regulatory compliance.
5. Higher tenant retention through tenant engagement programmes.
6. Risk mitigation through addressing climate risk and enhancing resilience.
7. In addition, academic research suggests that GRESB participation is a predictor of fund returns in terms of price appreciation but not income which is independent of local economic conditions (GRESB, n.d.; Devine et al., 2022).

NABERS ratings, which will be discussed later along with green building certifications, are already widely used in Australian commercial property and extending globally. Savills Australia reported on the positive impact of NABERS ratings on shopping centres making the following 7 observations:

1. Operating efficiency and reduced costs

LED lighting, modern HVAC systems with power factor correction units and on-site renewable energy generation result in lower energy costs with utility bills cut by up to 25%.

A NABERS rating can be helpful in helping building owners and managers to identify savings potential.

Regular meter audits contribute to reducing waste by analysis of energy consumption patterns. There may also be government grants for upgrades such as LED lighting.

2. Attracting Tenants and Shoppers

Consumers may favour businesses committed to or aligned with their or their personal or organisation's ESG values and practices. A high NABERS rating helps environmentally conscientious tenants and shoppers identity where to take their custom.

Retail tenants consider green building certification as part of their selection criteria. A high NABERS rating can boost occupancy rates and tenant retention.

3. Indoor Environment

Indoor Environmental Quality (IEQ) looks at air quality, lighting, and thermal comfort. Shopping centres that invest in these can see an improved shopping experience for shoppers and employees areas has a direct impact on both visitors and employees, improving their overall shopping experience, as well as their well-being. Better ventilation, reduced noise pollution and comfortable temperatures make the indoor environment more enjoyable and increase dwell times, increasing sales for tenants and increasing customer loyalty.

4. Environmental Leadership

Shopping centres that achieve high NABERS ratings actively support the national net zero by 2050 goal.

High NABERS ratings in carbon emissions, energy, water and carbon emissions demonstrate environmental leadership. This encourages other businesses, including tenants and suppliers, to do the same.

5. Future-Proofing and Regulatory Compliance

Australia, like other countries, is steadily moving towards stricter sustainability regulations. NABERS ratings can be expected to become a mandatory requirement. Shopping centres can stay ahead of the regulations and be well placed for future policy requirements.

A strong NABERS rating provides ESG conscious investors and buyers with reassurance that the property is sustainably managed and aligned with the requirements of environmental custodianship.

6. Enhanced Reputation and Marketability

Centres with top sustainability ratings can leverage the branding opportunity to be distinctive in their marketing campaigns, social media and public relations. This can enable them to attract eco-conscious shoppers, enhance brand image and visibility and strengthen market positioning.

In a similar way, this can result in increased appeal and support a green rent premium, and investors will see the rating as a sign that the property is well-managed.

7. Property Value and Investor Appeal

With increasing demand for green buildings, investors may pay a premium for properties with higher environmental performance. This results in future-proofed values, long-term competitiveness and attractive long-term investment assets.

Investing in sustainable practices helps meet the expectations of today's consumers, tenants, and investors. Using NABERS ratings as a tool enables shopping centres to improve their sustainability performance and gain a competitive edge (Hoban, 2024).

Sustainability, Capital Values and Resilience

A report published by JLL in 2023 covering the London office market using hedonic analysis methodology to isolate sustainability factors identified a premium for buildings with both BREEAM and EPC certifications. "Capital values were on average 20.6% higher as a result of BREEAM certification, with a single step improvement in EPC ratings producing a corresponding premium of 3.7%. Similarly, the average increase in rents associated with BREEAM certificates and a step improvement in EPC were 11.6% and 4.2% respectively" (Blake, 2023).

Sunny Zhang is based in Beijing. She is Chair of the Building Owners and Managers Association (BOMA) China Research Committee and Standing Director of the International Federation of Finance & Real Estate (IFFRE). She offered the following, which is worth sharing in full: "From an investment perspective, the positive correlation between ESG performance and financial returns has been fully verified. By deeply integrating BOMA operation standards with international sustainable development rating systems such as GRESB, commercial real estate can achieve the quantitative implementation of ESG value throughout the entire life cycle. For instance, CITIC Tower Beijing obtained the BOMA

Certificate of Excellence (COE) certification in 2023. Through dual-dimensional management and control of equipment optimization and intelligent operations, it achieved an 11.2% reduction in energy costs per unit area and a 15.4% decrease in carbon emission intensity. Such verifiable improvement in ESG performance has enabled it to gain significant advantages in financing costs and asset liquidity. For the green finance market, such ESG data on the operational side is precisely the core basis for solving the "greenwashing" problem and establishing a risk pricing mechanism".

CBRE's Hannah Jeong mentioned the impact different valuation methodology may play in allowing for ESG factors. As discussed, the DCF approach makes explicit recognition of rent and occupancy premiums and operating expense (OPEX) savings. She explained the importance of the capitalisation rate (cap rate), "a lower cap rate for green buildings means a higher valuation. The challenge is finding the market evidence to support the lower cap rate, for example if the cap rate for Hong Kong overall is 4%, the cap rate for a green building may be 3.5 to 4%. But valuers are reluctant to apply this because there is no recent market evidence to support it, many recent transactions are distressed sales. So much depends on the valuers judgement". With revenue and OPEX inputs both supported by evidence, Jeong adds "it is not double counting to reflect ESG factors in the cap rate as well because the cap rate reflects future risk".

Compared to DCF, Jeong points out that "market or comparable approaches tend to have less allowance for ESG factors with location, age, quality and size being the traditional key determinants. ESG factors are allowed for in the quality assessment but there is no formula and no differentiation for the level of certification because it is difficult to find evidence". Iris Kampers suggested that substantiated claims can be made around energy labels in the EU and that costs of getting a rating can be calculated and linked to rent and value increases. In some high-demand sectors, such as logistics, it may not be so clear-cut due to the lack of choice in the market.

The title of the JLL APAC Investor Sustainability Survey 2025, "Seizing the Green Opportunity Amidst Uncertainty", says it all. With all the various challenges in the market in 2025, improving sustainability performance supports enhanced returns and mitigates climate-related risks; this is "reshaping investment strategies and asset values across APAC".40% of respondents increased their offer based on sustainability considerations and only 9% said that sustainability considerations had not impacted investment decisions. 47% mentioned the financial viability of retrofitting as the biggest sustainability consideration. Expectation of new regulations and policies are high on the list of drivers within the next 5 years (Foo & Miglani, 2025).

In September 2025, ESG Today reported a statement from a European pension fund, PFZW, "has been developing a new investment strategy where financial performance, risk and sustainability are weighed equally within the framework of a total portfolio approach". According to the report, PFZW, did not renew their contracts with Blackrock and LGIM worth €14 billion and €15 billion respectively (Segal, 2025).

Does Sustainability Improve Resilience?

Resilience is a word that comes up a lot in the context of future proofing building value, both for new and existing buildings. The term encompasses design that accounts for current and future climate conditions, meaning sea level increases, extreme temperature and weather events. For existing buildings, it means adapting them. In both cases, resilient buildings can expect lower operating costs, improved occupier comfort and future-proofed value (Marquez, n.d.). The CBRE UK sustainability report notes that 6% of a sample of assets were at risk of river flooding at the end of 2024 (CBRE, 2024). One way to understand the value of resilience and risk to buildings can be seen is through insurance premiums. Building insurers have to evaluate future risks and price premiums accordingly and they have more experience in analysing and pricing risk than anyone else. According to CBRE, some building insurers are already withdrawing from US states including California and Florida (Mowell et al., 2023).

The National Council of Real Estate Investment Fiduciaries (NCREIF) is a US institutional real estate association founded in 1982; its data on US insurance costs as a share of operating expenses suggests they were 6% over the whole country in 2023, compared with 2% in 2000. In the states of Florida and California, the increases have been from just over 2% to nearly 12% and 5% to over 10%, respectively, over the same period. More recent climate-related events can only result in even higher insurance costs. Iris Kampers painted an even bleaker picture with the risk of heat stress or flood risk in some city centres potentially making them no go areas.

Does this mean there is a risk of an exodus of businesses and well-heeled residents from less resilient properties in cities and towns? This could lead to a repeat of the death of downtown seen in the 1980s when economic decline and increased crime resulted in widespread loss of value. This time climate risks leading to death of downtown v2.0 may be followed by resilience-led regeneration.

How realistic is this? In 2023, there were US$380 billion in total losses from natural catastrophes globally and US$118 billion in insured losses (PwC US, n.d.). The countries most impacted by natural disasters are China, the Philippines and the US (Marquez, n.d.) while according to insurer Munich RE, natural catastrophe events in 2019 caused damage of US$150 billion of which only US$52 billion was insured. Of the total annual damages, the US generally accounts for 30 to 35%, depending on the severity of the Atlantic hurricane season 80% (Moorcraft, n.d.). At the same time, 80% of US homeowners are not insured according to Munich RE.

The Insurance Council of Australia (ICA) is the industry association for the general insurance industry in Australia. According to the ICA Insurance Catastrophe Resilience Report 2024, 1.2 million Australian homes are at risk of flooding and the costs of extreme weather are predicted to more than double between 2021 and 2050. The cost of home insurance increased by 14% between 2022 and 2023 (Richardson & Bennett, 2024). On the other side of the planet, the Association of British Insurers recorded a 28% increase in weather-related claims for damage to homes and possessions in 2024 compared to 2023 (ABI, 2025).

A discussion with Andrew Slevin, Director of Charterfields a firm of Chartered Surveyors specialising in reinstatement valuations for the insurance industry, reinstatement costs in the event of a loss are based on the latest building code regulations which include ESG requirements. They do not include "betterment" that is a material improvement between the original building and a new best in class building. The Charterfields 2025 underinsurance report reveals "declared values for property damage policies are under declared in over 80% of locations, with some 37% of locations insured for 50% or less of the true reinstatement cost". According to Slevin, "building owners really need to consider the detail of what they will expect rather than simply indexing values or they risk missing out of the opportunity to make the best of an unhappy business that may be left in ruins after a storm or similar event". Reinstatement costs should include demolition, waste disposal, professional fees, building costs to meet latest codes and updated construction costs. This may be very different to the building value. This suggests a higher value is expected from a more sustainable building to replace the older one.

The following fuller explanation was shared by Andrew Slevin in relation to reinstatement valuations for insurance purposes:

1. "Insurers expect property owners to declare values that represent the full cost of reinstating the property 'as is'. In practice, insurance policies include wording that allow reinstatement to meet the latest building codes. This may be through a specific additional amount or percentage increase on the declared values mentioned in the insurance policy. While this reflects changes to code such as increased fire standards or changes to approved materials, building codes do not reflect what may be required from a market perspective to meet current occupier expectations. This means that, meeting the latest building code may not include for the increased energy efficiency, air quality standards and operational efficiencies that occupiers would expect. Therefore, in the event of an insured loss, owners or financiers may need to find additional funds beyond the reinstatement cost of the existing building to meet market expectations".
2. "While most buildings are insured, underinsurance is prevalent across the world and this is likely to put owners/funders in a difficult situation should a loss occur, namely they may not get a full settlement from insurers. This is likely to make it harder for owners to meet current costs to reinstate, never mind upgrade facilities to higher ESG standards".
3. "As has been seen with the dispute over the planned demolition for the M&S store on Oxford Road, London, there is a drive to repurpose existing buildings, to reduce the overall carbon footprint of a redevelopment, rather than just demolish and rebuild. However, this drive may conflict with what insurers need since they wish to limit business interruption losses. So, in the event a business has a significant insured loss, an insurer may be strongly opposed to any additional time and costs involved in trying to repurpose a structure since they want the business to be fully operational as fast as possible".
4. "Local and national governments are increasing the costs to remove and dispose of waste from sites, particularly those going to landfill.

Demolition costs are rising rapidly as a result. On a positive note, this can encourage the repurposing of buildings and the recycling of materials, but the additional costs (of either higher debris removal or repurposing/recycling) is not often understood by property owners, and therefore not being reflected in their declared values to insurers (which should include these costs)" (Slevin, 2025).

Brown Discounts: The Cost of Business as Usual

A brown discount may arise when buildings attract lower rents as a result of not being green and thus running the risk of becoming a stranded asset, that is, an asset that loses value due to being carbon intensive. According to the RICS, 50% of respondents to one of their surveys reported a brown discount in buildings not certified as green.

Real estate, with a few exceptions, is a longer term investment. It is not unknown for homeowners to live in the same house for 50 years or more. Most commercial property investors have a minimum hold period measured in years compared to minutes, or even seconds, for some financial products. In the case of the Crown Estate and family investment trusts, hold periods may add up to centuries. Homes generally undergo periodic redecoration, insulation improvements, installation of double glazing, solar power installation, hot water boiler efficiency upgrades and so on as part of a natural home improvement. It is generally accepted that the cost to the owner of such improvements may not be fully reflected in the valuation but they may improve saleability and help maintain valuation in line with the general market. Those that do not will be expected to sell for a price that reflects the cost of upgrading.

This is a view confirmed by Jeong, "the cost of retrofitting a property to meet sustainability standards will almost invariably be deducted from a transaction price". Kampers also referred to this "the 15% of worst performing buildings in the Republic of Ireland will need to be renovated by 2030 and 26% by 2033. This means that hotels with an F rated EPC will have to be A + + when 2 years ago they didn't even need an EPC at all". That's not all, there's a massive data gap, "Germany and Spain don't yet even have a database of all energy labels". Coincidentally, the Dutch Green Building Council office is in a building, Van Bylandt Huis, that does not appear on the Dutch database of green buildings, despite having been "stripped completely and was built from the ground up" in a 1999 renovation and again refurbished again in 2014, according to the property description (CBRE, n.d.).

A similar principle applies to commercial property when owners fail to understand transition risk that can result in loss of tenants, exposure to carbon pricing, missing out on grants to upgrade and reputation risk. According to Carbon Risk in Real Estate Monitor (CRREM), buildings that fail to keep up with regulatory efficiency standards and/or market demand risk economic obsolescence, become less marketable and require expensive retrofitting may become "stranded assets" (CRREM, n.d.). A resulting loss of value will be of concern not only to the owner but to their shareholders, bankers, insurers and other stakeholders. Banks may refuse to lend to buyers of such assets; insurers may demand higher premiums. In extreme cases, the cost of retrofitting may exceed the value of the property.

The cost of doing nothing may not be immediately apparent. Lizette McNeill, Managing Partner at IXO Partners, a specialist climate and engineering firm in Hong Kong offers a warning to new projects, "buildings being designed and built today are being built to today's standards. The problem with this is that today's standards do not meet current changing environmental conditions including both the gradual changes to heat, rain intensity, humidity, and catastrophic events such as typhoons, intense rain storms, flood and fire risk as well as seismic activity". In the United Kingdom, 2025 saw the second hottest June and fifth hottest July since 1885 but the future homes standard, due to be published, has a focus on insulation to keep the heat in and little for homes overheating in the summer (Gecsoyler & Goodier, 2025).

The Business Case for Net Zero Buildings

While surveys suggest most respondents want society to balance emissions, the fierce political debate on the relative merits and demerits of net zero remains. It can be easy to forget the fundamentals. Net zero reduces dependency on the price fluctuations of fossil fuels and therefore includes a considerable risk and cost avoidance element. This is set out in detail by BOMA Canada in "Lessons from the Net Zero Challenge", a programme to recognise buildings that showcase sustainability in design and operations. According to their report, the business case includes:

1. Increased revenues from higher rents and occupancy rates.

2. Future proofing values.
3. Lower operating costs.
4. Carbon pricing /emission cap avoidance.
5. Reduced climate risk.
6. Building code compliance.
7. Access to grants and incentives.
8. Lower-interest financing (BOMA Canada, 2022).

The above notwithstanding, as Kampers stated above, there is little or no evidence that among the lower emission buildings, net zero buildings are not yet presenting a more compelling business case compared to an A-rated building.

Case studies that provide additional support for the business case are included in Chapter 10, The Changing Landscape.

Summary

The business case for sustainable built environments should now be clear. For building owners, the imperative is part financial and part resilience driven. For infrastructure assets such as airports, railways and roads, there are obvious financial risks. Business interruption for an airport due to flooding, for example, is a serious financial loss. 100 existing airports are below mean sea level and 1,238 are in the Low Elevation Coastal Zone (Yesudian & Dawson, 2021). The case for risk avoidance for buildings and infrastructure is easily understood.

What is less certain is the precise relationship between green buildings and their value. Some property developers, investors and fund managers have shared that they are not very keen on green building certification and similar sustainability tools. There were no conversations that suggest they do not like lower operating costs, lower risks, rent premiums and capital values for their properties. A change of mindset is required. The kind of granular value adding metrics that investment analysts are used to poring over in the due diligence process are not always available and irrefutable. This lack of clear data leaves ESG factor valuations open to challenge, a point made by the interviewees. A great deal rests on the shoulders of valuation professionals and investment analysts. The judgement of valuers now has an additional burden, assessing future climate-related risk for a specific asset in today's values. Availability of transparent market evidence will help the defence of green building valuations.

References

ABI. (2025, February 10). *More action needed to protect properties as adverse weather takes record toll on insurance claims in 2024*. ABI. Retrieved April 23, 2025, from https://www.abi.org.uk/news/news-articles/2025/2/more-action-needed-to-protect-properties-as-adverse-weather-takes-record-toll-on-insurance-claims-in-2024/

Aronsohn, A. (2024, March 28). *Beyond Numbers: Incorporating ESG into International Valuation Standards*. IVSC. Retrieved February 15, 2025, from https://ivsc.org/esg2024/

Blake, K. (2023, January 17). *Environmentally sustainable real estate attracts higher prices*. Retrieved August 23, 2025, from https://www.jll.com/en-uk/newsroom/environmentally-sustainable-real-estate-attracts-higher-prices

BOMA Canada. (2022). *Lessons from the Net Zero Challenge*. Retrieved February 21, 2025, from https://bomacanada.ca/wp-content/uploads/2023/04/Net-Zero-ReportFIN.pdf

CBRE. (2023, March 16). *The green building premium: Does it exist*. Retrieved February 27, 2025, from https://www.cbre.com/insights/viewpoints/asia-pacific-viewpoint-the-green-building-premium-does-it-exist

CBRE. (2023, February). *Strengthening value through ESG*. Retrieved February 12, 2025, from https://www.cbre.com/insights/books/strengthening-value-through-esg/value-of-environmental-building-features

CBRE. (2024). *CBRE UK sustainability index*. Retrieved March 27, 2025, from https://mktgdocs.cbre.com/2299/5349bb8d-a739-4cba-9451-0e3c458dfd15-908628925/UK_Sustainability_Index_Result.pdf

CBRE. (n.d.). *Van Bylandthuys, Benoordenhoutseweg 46*. Retrieved February 18, 2025, from https://www.cbre.nl/en-gb/properties/office/details/NL-4615/van-bylandthuys-benoordenhoutseweg-46-2596-bc-s-gravenhage-s-gravenhage-2596-bc

Chen, Y. (2023, December 7). *Green premium study of New York and London Real Estate finds strong evidence for a 'green premium'*. Retrieved February 15, 2025, from https://www.ubs.com/au/en/assetmanagement/insights/asset-class-perspectives/private-markets/articles/green-premium.html

Coleman, R. (2024, September 12). *Sustainable Warehouses Command 19% price premium underlining 'flight to quality' in European Logistics Investment*. Retrieved February 12, 2025, from https://www.cushmanwakefield.com/en/united-kingdom/news/2024/09/sustainable-warehouses-command-19percent-price-premium

CRREM. (n.d.). *CRREM initiative's definition on stranding risk and stranded assets in the build environment what is stranding risk and how is the term related to transition risk?* CRREM. Retrieved April 23, 2025, from https://www.crrem.eu/wp-content/uploads/2022/12/CRREM-initiative-definition-on-stranding-risk-and-stranded-assets-in-the-build-environment.pdf

Devine, A. Sanderford, A., & Wang, C. (2022, June 16). Sustainability and private equity real estate returns. *The Journal of Real Estate Finance and Economics*, *68*. https://doi.org/10.1007/s11146-022-09914-z

Foo, Y. L., & Miglani, K. (2025). *Seizing the green opportunity amidst uncertainty*. JLL. Retrieved September 17, 2025, from https://www.jll.com/en-au/insights/seizing-the-green-opportunity-amidst-uncertainty

Ganz, E. (2024, June 12). *Embracing the future: approaches to ESG in real estate valuation*. EY. Retrieved April 23, 2025, from https://www.ey.com/en_ch/insights/real-estate-hospitality-construction/embracing-the-future-approaches-to-esg-in-real-estate-valuation

Gecsoyler, S., & Goodier, M. (2025, August 12). 'Do not buy these flats': residents warn about unbearable heat inside London new-builds. Retrieved August 12, 2025, from https://www.theguardian.com/australianews/2025/aug/12/do-not-buy-these-flats-residents-warn-about-unbearable-heat-inside-london-new-builds

GRESB. (n.d.). *Key financial benefits of GRESB participation*. Retrieved February 15, 2025, from https://www.gresb.com/nl-en/key-financial-benefits-of-gresb-participation/

Hoban, L. (2024, December 19). *The positive impact of NABERS ratings on shopping centres in Australia*. Retrieved February 15, 2025, from https://www.savills.co.jp/blog/article/219578/australia-articles/the-positive-impact-of-nabers-ratings-on-shopping-centres-in-australia.aspx

JLL. (2022, November 29). *The value of sustainability: Evidence for a green premium in Asia*. Retrieved February 12, 2025, from https://www.jll.com.hk/en/trends-and-insights/research/the-value-of-sustainability-asia-pacifics-green-premium-opportunity

Marquez, T. (n.d.). *Climate-resilient buildings are the way forward for finance, architects, and engineers*. World Green Building Council. Retrieved February 15, 2025, from https://worldgbc.org/article/climate-resilient-buildings-are-the-way-forward/

Moorcraft, B. (n.d.). Natural Catastrophe Report. *Insurance Business America*. https://issuu.com/keymedia/docs/iba8.08_lr_ipad?fr=sODJiMTE3ODU1MzM

Mowell, M., Leahy, M., & Schoenmaker, D. (2023, September 8). *Insurance costs are no longer worry-free*. Retrieved February 15, 2025, from https://www.cbre.com/insights/briefs/cotw-insurance-costs-are-no-longer-worry-free

Onishi, J., Deng, Y., & Shimizu, C. (2021, November 5). Green premium in the Tokyo Office rent market. *Sustainability*, *13*(21). https://doi.org/10.3390/su132112227

PwC US. (n.d.). *Building climate resilience: Insurers can guide the way*. Retrieved February 17, 2025, from https://www.pwc.com/us/en/industries/financial-services/library/climate-risk-and-insurance.html

Richardson, D., & Bennett, E. (2024, November 4). *Premium price: The impact of climate change on insurance costs*. The Australia Institute. Retrieved April 23, 2025, from https://australiainstitute.org.au/report/premium-price-the-impact-of-climate-change-on-insurance-costs/

RICS. (2024). *About RICS*. RICS. Retrieved May 29, 2024, from https://www.rics.org/about-rics

Savills. (2022, June 1). *Is There a Green Premium, or a Brown Discount? | Themes*. Savills. Retrieved May 31, 2023, from https://www.savills.com/prospects/themes-is-there-a-green-premium-or-a-brown-discount.html

Scherrenberg, J., Wessels, P., & Nelisse, P. (2024, January). *The future of real estate valuations: the impact of ESG*. Retrieved February 4, 2025, from https://www.rics.org/content/dam/ricsglobal/documents/latest-news/WBEF-ESG-and-valuation-2024.pdf

Segal, M. (2025, September 4). *BlackRock, LGIM Lose $34 billion in mandates from Dutch Pension Fund's shift to sustainability & active-focused investment policy*. ESG Today. Retrieved September 17, 2025, from https://www.esgtoday.com/blackrock-lgim-lose-34-billion-in-mandates-from-dutch-pension-funds-shift-to-sustainability-focused-investment-policy/

Ulbrich, C. (2022, January 12). *How the conversation around green real estate is changing*. The World Economic Forum. Retrieved February 23, 2025, from https://www.weforum.org/stories/2022/01/green-real-estate-sustainability-corporate-priority/

Wong, M., Leung, L., Lam, M., Cheung, J., Kong, E., & Wan, W. (2024, May). *Sustainable assets, sustainable returns: Unveiling the ESG-rental premiums nexus*. Retrieved February 14, 2025, from https://content.knightfrank.com/research/2837/documents/en/hong-kong-esg-report-2024-11222.pdf

Yesudian, A., & Dawson, R. (2021). Global analysis of sea level rise risk to airports. *Climate Risk Management*, *31*. https://doi.org/10.1016/j.crm.2020.100266

CHAPTER 3

What Is Sustainability in the Built Environment?

"*We shape our buildings; thereafter they shape us*". Sir Winston Churchill, October 28, 1943, requesting that the House of Commons bombed out in May 1941 be rebuilt exactly as before.

What Is a Sustainable Built Environment?

The built environment includes all activities, processes and decisions in the making of environments for human activities and occupants. This includes buildings, construction and roads, highways and other infrastructure across the entire life cycle; it is the antithesis of sustainability. According to a 2023 United Nations Environment Programme (UNEP) report "The global built environment sector is by far the largest emitter of greenhouse gases, responsible for at least 37% of the global emissions" (UNEP, & Yale Center for Ecosystems + Architecture, 2023). Other sources may give slightly different figures and the exact figure may vary by year. For example, the buildings and construction sector combined accounted for 39% of total global energy-related CO2 emissions in 2018, 38% in 2019 (UNEP, 2020), and 34% in 2023 (Hamilton & Hsu, 2025). Elsewhere, 30% of global energy emissions are related to building operations, directly and indirectly (IEA, 2022a, 2022b). Regardless of the exact figures, it is clear that the industry sucks in vast quantities of resources

S. Crispin, *Sustainability in the Built Environment*, https://doi.org/10.1007/978-981-95-6549-8_3

in the form of materials (extraction, processing, manufacture, transportation, construction and disposal related energy) and labour like no other. The European Commission estimates that construction in cities accounts for around 50% of all extracted materials used, 30% of water used and generates 33% of waste. The entire sector, which we refer to as the built environment, directly and indirectly contributes more man-made GHG emissions across its entire life cycle than any other human activity.

To make matters worse, buildings, cities and transportation systems are the bedrock on which further resource-intensive and environmentally destructive economic activity is built. Yet cities are essential to provide for the basic needs of humanity today and the most efficient means of ensuring continued prosperity.

Environmental topics such as bio diversity loss, water and air pollution, plastic waste, deforestation, renewable energy, agriculture and transport all get a lot of media coverage. Little attention is given to the built environment itself, but sustainable built environments would go a long way to tackling most of these problems. Wouldn't a top-down approach be more effective?

What exactly is the built environment? We have had one definition above but there are others. It is not an easy term to define and one that is relatively new; "any environment which is not natural" might be one simple answer. UNESCO defines it as "the human-made environment that provides the setting for human activity, ranging in scale from buildings to cities and beyond" (UNESCO, n.d.). This includes the construction and use of buildings of all sizes, cities and city regions, infrastructure and other related man-made structures of all kinds. This includes everything from the ancient pyramids of Giza to the International Space Station. My own alternative offering is "all man-made structures overlaid on the natural environment".

For the purposes of this book, we will focus on the buildings and construction sector with the occasional foray into infrastructure but without dwelling on man-made structures outside the Earth's atmosphere. When I have a building in mind, it is a 50 or so storey, steel and concrete structure with a glass façade, mixed use commercial building completed in the mid-1980s in a non-core location on Hong Kong Island.

How on earth can such structures be sustainable? To answer that, we need to understand exactly what is meant by sustainable. With increasing concern among government bodies the World Commission on Environment and Development (WCED) was convened by the United Nations in 1983 and in 1987 published the Brundtland Report (Thomsen, 2013). It set out many of the guiding principles for sustainability that we refer to today and defined sustainable development as "development that meets the needs of the present without compromising the ability of future generations to meet their own needs" (World Commission on Environment and Development, 1987). This is an important definition; the whole idea of sustainable development means not compromising the needs of future generations. The allocation of resources to different communities and geographies is an inherent part of sustainability.

Another important definition to clarify at this point is climate change. Climate change "refers to long-term shifts in temperatures and weather patterns. These shifts may be natural but since the 1800s, human activities have been the main driver of climate change, primarily due to the burning of fossil fuels (like coal, oil and gas) which produced heat-trapping gases" (Kolawole & Okonkwo, 2022).

We Are Using More Than Can Be Regenerated

The modern, urbanised world we inhabit is not sustainable. Earth Overshoot Day is calculated by Global Footprint Network and marks the date in the calendar when our collective use of natural resources exceeds what Planet Earth can regenerate that same year. In 1971, it was in December; in 2023 it was in August (Global Footprint Network, n.d.). Different countries have different Overshoot Days. While that of the United States falls in March, that of Jamaica is not until the 2nd half of December.

The measure used is global hectares (gha). Different countries have a different gha as well as a different global biocapacity, the capacity of an ecosystem to regenerate biological materials used by people and absorb man-made waste. At current levels of consumption, if everyone lived like the USA, we would need 5.1 planets. The equivalent for India is 0.8. Globally we need 1.75 planets to be sustainable (Global Footprint Network, n.d.). Clearly this is not possible.

We need sustainable built environments that can be used today without depleting resources from the needs of future generations. That means being resource and energy efficient, reusable and recyclable. This is known as decoupling, where economic growth is able to take place without additional environmental harm.

The problem is that none of these is in any way easy. We will consider the inputs in the next chapter. But needless to say, building materials including concrete, steel and glass as well as energy used for heating, air conditioning, lighting and lifts are all energy intensive; while the exact number changes year-on-year, they account for 34–38% of total energy-related CO2 emissions. We have a great challenge on our hands.

There are institutions trying to improve things. The World Green Building Council (WGBC) is a global action network to accelerate sustainability in the built environment; it seeks to guide the building and construction sector towards a sustainable transition through 4 impact areas:

1. Climate action, meaning decarbonisation of the built environment.
2. Health and well-being, meaning "healthy, equitable and resilient buildings, communities and cities".
3. Resources and circularity, meaning regeneration of resources and natural systems as well as a circular economy providing jobs and economic benefit.
4. Sustainable finance and ESG (WGBC, n.d.).

WGBC makes the point that a sustainable building may be different depending on the local climate, building age and type among other things. The general principles are:

1. Energy efficiency.
2. Net zero whole life carbon.
3. Adaptation to climate change.
4. Renewable energy powered and all electric.
5. Support health and well-being.
6. Making a positive social impact across the supply chain.
7. Supporting a just transition towards green jobs and use of sustainable finance.
8. Use the circular economy to minimise material use and waste.
9. Regenerate nature, especially focus on water resources.
10. Take a systems approach by including infrastructure, community and city level action to promote a sustainable business model (WGBC, n.d.).

The Increasing Scale of the Challenge

Another challenge is that the scale of the built environment is increasing. The Global Cement and Concrete Association state that 75% of the infrastructure that will exist in 2050 has not been built yet (GCCA, 2021) while the International Energy Agency (IEA) has suggested the total floor area of buildings will increase from 244 billion sq.m. in 2020 to 427 billion by 2050 (IEA, 2022a, 2022b). Nearly 90 billion sq.m. of new building stock is expected in Africa, 50 billion sq.m in the People's Republic of China (PRC) and 45 billion in India by 2060 (Zhou et al., n.d.), a total just over the entire land area of Cambodia. To understand the scale of the challenge, around 60% of the buildings that will exist in India by 2040 do not yet exist and buildings in ASEAN, the PRC and India account for 27% of the global total building-related energy consumption (Zhou et al., n.d.). Cities in sub-Saharan Africa housed 472 million people in 2018 and this is expected to double before 2050 (Saghir & Santoro, 2018).

In terms of economic activity, the gravitational pull to Asia is also strong. According to a report published by Savills with data from Oxford Economics, Tokyo, Osaka, Nagoya and Seoul are among the top 10 global city economies. Shanghai, Beijing, Shenzhen and Delhi are forecast to enter the top 10 by 2050. With economic growth comes increased consumption and a corresponding increase in related emissions (Savills, 2024).

It is not just about new buildings; at least they have the opportunity to be sustainable from the ground up. According to various IEA reports, at least 40% of existing buildings in developed economies predate the first thermal regulations in 1980 (IEA, 2022a, 2022b). The problem with older buildings is that, while shorter lifespan mechanical installations can be upgraded as they reach replacement dates, the structure and 'building envelope' are rarely upgraded but, if built with low carbon materials, can have the greatest impact on reducing CO2 emissions (IEA, 2022a, 2022b). Current retrofitting rates are only 1%, many of which are shallow retrofits; this compares to the 2022-quoted IEA figure of 20% deep retrofit to a zero-carbon-ready level by 2030. It is imperative that sustainable building technology and methods are adopted in all urbanising markets and that already urbanised markets upgrade their existing building stock to optimise energy and resource efficiency.

Evolution of ESG: Focus on the E

At this point, it might be useful to pause from the breathlessness of the state we find ourselves in and look at the evolution of Environmental, Social and Governance (ESG) reporting in business. In particular, let us consider the role ESG plays in navigating our way through decarbonising the built environment and the business case mentioned in the previous chapter.

Don't hold your breath. In July 2022, the front page of The Economist declared "ESG, 3 letters that are not going to save the planet". This came after an apparent wave of adoption of ESG reporting among listed companies worldwide, covering non-financial metrics classified as Environmental, Social or Governance topics. At the time of The Economist publication, according to accountancy firm KPMG, nearly all the top 250 companies globally reported on sustainability (KPMG, n.d.). It is a matter of controversy among businesses and politicians alike. In June 2024, the Financial Times headline "The unsustainable hype around ESG" topped a story about the outflow of funds from ESG investments due to underperformance and politicisation. A telling quote "ESG is something of a Frankenstein's monster of buzzwords" and perhaps a cover for greenwashing. Also notable, "reaching net zero is seen as good business by executives. Mitigating governance issues is viewed as necessary risk management. Green investments are increasingly profitable" (The Editorial Board, 2024).

The term resilience has gained traction as a result of the geopolitical pushback against ESG in 2024 and 2025. It is a term that wraps sustainability into a climate change adaptation and unstable weather mitigation mode (Khan, 2025). Resilience signals a risk management response to uncertainty of a changing climate regardless of the cause, including the perception of political risk that comes with being too openly ESG.

The Freshfields Report "Who Cares Wins" formed part of the United Nations 2006 Principles for Responsible Investment (PRI) and set out the requirement for ESG (Environment, Social Governance) factors to be incorporated into the evaluation of companies alongside financial reporting (Atkins, 2020). Following the Paris Agreement in 2015, by 2017 shareholder activism was pushing for progress on ESG factors. According to global law firm Hogan Lovells, shareholder activists in the US use annual general meetings (AGMs) to put non-binding proposals to shareholder vote. In the UK, they may also mount public campaigns on key issues and even court cases on ESG issues. In this way, the risk of having to defend lawsuits for lack of, incomplete or misleading ESG statements and claims is a driver for implementation and transparent reporting. This may be referred to as greenwashing. We also see a pushback in the form of anti-ESG activism; this may result in the less well-known 'greenhushing' where a company downplays or understates its ESG credentials. This may also include deliberately setting soft goals and targets that are easily achieved (Gartner et al., n.d.). Is it time to focus efforts on the E while the S and the G go into a holding pattern?

Looking further back, it is possible to link ESG to the earlier lens of Elkington's triple bottom line (TBL) which considers people, planet, profit reporting simultaneously on social (people), environment (planet) and financials (profit) (Grand Canyon University, 2021). Such reporting was intended as a means for a business to present the total cost of its operations, including externalities, in other words looking at more than very narrow financial terms. These, with financials already covered by existing accounts, are the core elements of ESG reporting.

25 years later, Elkington effectively "recalled the product" on the grounds that none of the sustainability frameworks went far or fast enough "to stop us all overshooting our planetary boundaries" (Elkington, 2018). This thinking has gained traction, as suggested by The Economist, but abiding by ESG principles is good business, as stated by the Financial Times. The challenges for the built environment are even more immense than this debate, as we will explore in the following chapters.

Regardless ESG reporting, distinct from ESG investing, is now the established framework by which businesses report or are required to report by stock exchange listing rules on non-financial performance; they are able to demonstrate their actions to decarbonise, the support they give to the communities in which they operate (a more strategic and measured approach to Corporate Social Responsibility) and their compliance with legal, tax and other regulations. It is a positive that so many commitments are being made by so many organisations and that shareholders have some transparency about business resilience in the face of climate-related risk but there is criticism that reporting and regulatory compliance divert attention from the urgent task of reducing Greenhouse Gas emissions.

Shareholder Activism

Major financial institutions have proxy voting and engagement guidelines. One example is State Street Global Advisors; their published North America Proxy Voting and Engagement Guidelines expect 30% women directors at board level for Russell 3000 companies, with the potential sanction of voting against all incumbent members after 3 years of non-compliance. They also expect "at least one director from an under-represented racial/ethnic community on its board" (Colton, 2023). This means that institutional investors may vote against the re-election of directors where they have not made progress in appointing more women to their boards. In another case in 2017, shareholders voted to require ExxonMobil to report on the impact of climate change on its business (Rushe, 2017).

ESG and the Built Environment

Considering how ESG applies to the built environment in particular, the environmental side means considering embodied and operating carbon. That is the carbon in the materials that go into the building and construction process and the carbon related to the operational stage of the building life cycle. On the social side, this includes the workers involved in mineral extraction, processing, material manufacturing, construction and all aspects of property management including cleaning of the exterior and the work environments for site management. On the Governance side, site acquisition and planning applications can be notorious areas where illicit payments or entertainment may creep in, compliance with building codes, tax, valuations and some of the less obvious may include tendering processes, financing and insurance. Land, buildings and construction-related businesses may seem never far from potential scandal and controversy leading to financial ruin.

Sunny Zhang offered her thoughts that ESG has "evolved from a policy-driven compliance requirement to a core dimension for financial markets to evaluate asset quality, and systematic practices on the operational side are the key to realizing ESG value. According to the *2024 BOMA China Commercial Building Tenant Satisfaction Benchmark*

Report, 100% of the surveyed organisations attached importance to environmental management, and 96% have listed energy efficiency improvement as a core demand. This shift in demand is directly transmitted to the asset value dimension".

Ground Net Zero: The Focus on Asia

With an estimated half of all global CO2 emissions generating in Asia, rapid urbanisation trends and energy hungry cooling requirements, Asia, among all the rapidly urbanising economies, is the particular focus of this book. According to the 2018 Revision of World Urbanisation Prospects published by the UN Department of Economic and Social Affairs (UN DESA), by 2050 68% of the world's population is expected to live in urban areas compared to 55% in 2018. This means an additional 2.5 billion people living in cities, of which 90% of the growth will be in Asia and Africa (UN DESPA, 2018).

According to UN-Habitat, 54% of the world's urban population live in Asia and ongoing urbanisation means that over 55% of Asia's population will be urban by 2030; cities are growing into each other to create city clusters or megalopolises such as the Greater Bay Area in Southern China, and more than 50% of the world's new buildings are built in Asia (USAID, n.d.). Asia is also the region where the cities are sinking, where population densities are increasing and where the climate demands energy hungry air conditioning in buildings.

The result of all the built environment expansion, according to the IEA is that the global total floor area of buildings will increase from 244 billion sq.m. in 2020 to 427 billion by 2050 (IEA, 2022a, 2022b).

The impact of increasing urbanisation is not just about new buildings. Existing buildings tend to be less efficient and less likely to be green certified. "According to the Hong Kong Green Building Council (HKGBC), there is a disparity of 1,700 new buildings and 100 existing buildings that have been green certified", said Jonathan Lee, Deputy Managing Director at Savills at an Australian Chamber of Commerce event in November 2023. To put this number in context, there are around 50,000 buildings in Hong Kong of which nearly 14,000 will be more than 50 years old by 2030 (Loh, 2023).

Urbanisation is accompanied by economic growth. Development of cities also results in improved provision of basic services, reductions in poverty as well as improved literacy and life expectancy. In 1988, Brazil went so far as to amend their Constitution to embed "the social function of cities" (UN Habitat, 2010).

Economic growth increases demand for electricity; as early as 2025, Asia is expected to account for 50% of global electricity consumption compared to 25% in 2000, with China alone consuming around 33% (Buchholz, 2023). Looking at the impact of urbanisation and energy use in China between 1980 and 2010, Zhao & Zhang (2018) showed that "for every 1% increase in the urban population relative to the national population, national energy consumption rose 1.4%".

A major driver of electricity consumption is the need for cooling in the hot and humid parts of Asia; this means that the 2020 estimate of 1.9 billion air conditioning units globally in 2020, (mainly in the US, China, Japan and South Korea), is expected to grow to 5.5 billion by 2050 and air conditioning-related energy demand set to triple to 6205 TWh (Armstrong, 2020). As reported by the BBC, India is the fastest growing air conditioning market, 93% of households rely on fans for comfort and figures from IEA suggest that by 2050 the potential total home air conditioning-related energy consumption in India could exceed that of the whole of Africa in 2024 (Biswas, 2024).

Rapid urbanisation is fuelling massive demand for energy hungry appliances. Taken together, this means that of the 30% global energy-related CO2 emissions from building operations, 50% is derived from Asia and 37% of that is air conditioning-related. In other words 5.5% of global energy-related CO2 emissions is derived from air conditioning in Asia (excluding refrigerants), as we will learn in Chapter 11 this is similar to the entire aviation and maritime industries combined.

This means that one of the critical factors in decarbonising the built environment is how efficiently buildings in Asia can manage to keep their occupants cool. This is not limited to new buildings; a particular focus should be on making existing buildings more efficient. 40% of the emissions reductions needed to meet Paris Agreement targets can be delivered by energy efficiency (Tveen, 2021). The good news, which we will get to, is that existing building operations are relatively easy to improve and new buildings can be efficient from the ground up.

Summary

Looking ahead, it is imperative that both new buildings are nearly net zero and that existing buildings are upgraded (retrofitted) to optimise energy and resource efficiency. This needs to be done with urgency and reporting requirements that allows organisations to focus efforts on emission reductions as an imperative.

Relative to other sectors like aviation, meat production and vehicles, we hear little about buildings and construction in the environmental conversation. If the industry is not talking about the problem, or working in silos, how can it address it? This book seeks to bridge that knowledge gap and start a process of understanding, acknowledgement and present ideas to solve the problem.

In the next two chapters we will look at why the built environment matters, the role it plays in all our lives and the resources that go into buildings, the building materials (land, cement, aggregate, steel, aluminium, glass, insulation, water, labour and finance) and the construction process. Then we will consider the various efforts at decarbonising materials from extraction to construction new materials and processes.

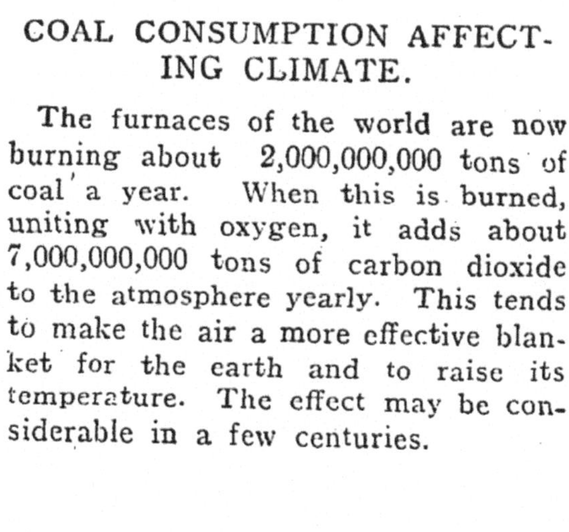

COAL CONSUMPTION AFFECTING CLIMATE.

The furnaces of the world are now burning about 2,000,000,000 tons of coal a year. When this is burned, uniting with oxygen, it adds about 7,000,000,000 tons of carbon dioxide to the atmosphere yearly. This tends to make the air a more effective blanket for the earth and to raise its temperature. The effect may be considerable in a few centuries.

Fig. 3.1 Coal consumption affecting climate (The Rodney and Otamatea Times, 1912, Page 7 licensed under Creative Commons BY-NC-SA 4.0)

Fig. 3.2 Natural environment to built environment, Victoria Road, Hong Kong (authors photograph)

Fig. 3.3 Front cover of The Economist (July, 2022)

References

Armstrong, M. (2020, August 27). *Air conditioning biggest factor in growing electricity demand*. Statista. Retrieved March 4, 2024, from https://www.statista.com/chart/14401/growing-demand-for-air-conditioning-and-energy/

Atkins, B. (2020). Demystifying ESG: Its history & current status. *Forbes*. https://www.forbes.com/sites/betsyatkins/2020/06/08/demystifying-esgits-history--current-status/?sh=6a4ebc2b2cdd

Biswas, S. (2024, June 27). *India heat: What record air conditioner sales reveal about heatwave*. BBC. Retrieved June 17, 2025, from https://www.bbc.com/news/articles/cw555ny3q9xo

Buchholz, K. (2023, February 15). *Asia will use half of the world's electricity by 2025*. The World Economic Forum. Retrieved March 4, 2024, from https://www.weforum.org/agenda/2023/02/electricity-demand-asia-energy/

Colton, B. (2023, May 8). *North America proxy voting and engagement guidelines*. The Harvard Law School Forum on Corporate Governance. Retrieved

May 29, 2024, from https://corpgov.law.harvard.edu/2023/05/08/north-america-proxy-voting-and-engagement-guidelines/

Elkington, J. (2018). 25 years ago I coined the phrase "triple bottom line." Here's why it's time to rethink it. *Harvard Business Review, 25*, 2–5. https://hbr.org/2018/06/25-years-ago-i-coined-the-phrase-triple-bottom-line-heres-why-im-giving-up-on-it

Gartner, O., Hamberger, M., Croke, A., Evans, N., Racine, K., Wuertz, A., & Regan, W. (n.d.). *Hogan Lovells*. Retrieved May 29, 2024, from https://www.engage.hoganlovells.com/knowledgeservices/news/recent-developments-in-esg-shareholder-activism-around-the-world-and-suggestions-for-risk-mitigation

GCCA. (2021, October 12). *Global cement and concrete industry announces roadmap to achieve groundbreaking 'Net Zero' Co2 emissions By 2050.* Retrieved August 23, 2025, from https://gccassociation.org/news/global-cement-and-concrete-industry-announces-roadmap-to-achieve-groundbreaking-net-zero-co2-emissions-by-2050/

Global Footprint Network. (n.d.). *About - Earth overshoot day*. Earth Overshoot Day. Retrieved May 29, 2024, from https://overshoot.footprintnetwork.org/about/

Grand Canyon University. (2021, July 30). *People, planet, profit: The Triple Bottom Line | GCU Blog*. Grand Canyon University. Retrieved May 29, 2024, from https://www.gcu.edu/blog/business-management/people-planet-profit-looking-triple-bottom-line

Hamilton, I., & Hsu, S. (2025, March 1). *Not just another brick in the wall.* GlobalABC. Retrieved June 17, 2025, from https://globalabc.org/sites/default/files/2025-03/Global-Status-Report-2024_2025.pdf

IEA. (2022a). Technology *and innovation pathways for zero-carbon-ready buildings by 2030*. Introduction – Technology and innovation pathways for zero-carbon-ready buildings by 2030 – Analysis - IEA. Retrieved May 29, 2024, from https://www.iea.org/reports/technology-and-innovation-pathways-for-zero-carbon-ready-buildings-by-2030/introduction

IEA. (2022b), *Renovation of near 20% of existing building stock to zero-carbon-ready by 2030 is ambitious but necessary*. IEA, Paris https://www.iea.org/reports/renovation-of-near-20-of-existing-building-stock-to-zero-carbon-ready-by-2030-is-ambitious-but-necessary

International Energy Agency. (2022, October 26). *Global buildings sector CO2 emissions and floor area in the Net Zero Scenario, 2020–2050 – Charts – Data & statistics*. IEA. Retrieved May 22, 2023, from https://www.iea.org/data-and-statistics/charts/global-buildings-sector-co2-emissions-and-floor-area-in-the-net-zero-scenario-2020-2050

Khan, Y. (2025, February 28). Don't call it ESG, call it resilience. *WSJ*. https://www.wsj.com/articles/dont-call-it-esg-call-it-resilience-067a2a9a

Kolawole, E., & Okonkwo, W.I. (2022). Impacts of climate change on environment and the remedies. *International Journal of Weather, Climate Change and Conservation Research*. Retrieved June 12, 2025, from https://www.eajournals.org/wp-content/uploads/Impacts-of-Climate-Change-on-Environment-and-the-Remedies.pdf

KPMG. (n.d.). *Key global trends in sustainability reporting - KPMG Global*. KPMG International. Retrieved March 1, 2024, from https://kpmg.com/xx/en/home/insights/2022/09/survey-of-sustainability-reporting-2022/global-trends.html

Loh, C. (2023, February 10). Aging buildings problem needs to be addressed aggressively. *China Daily*. https://www.chinadailyhk.com/article/a/314422

Rushe, D. (2017, May 31). Shareholders force ExxonMobil to come clean on cost of climate change. *The Guardian*. https://www.theguardian.com/business/2017/may/31/exxonmobil-climate-change-cost-shareholders

Saghir, J., & Santoro, J. (2018, April 12). *Urbanization in Sub-Saharan Africa*. Retrieved August 23, 2025, from https://www.csis.org/analysis/urbanization-sub-saharan-africa

Savills. (2024, February 22). *Booming cities will drive Asia Pacific growth*. Savills. Retrieved May 29, 2024, from https://www.savills.com/prospects/cities-booming-cities-will-drive-asia-pacific-growth.html

The Editorial Board. (2024, June 9). The unsustainable hype around ESG. *Financial Times*. https://www.ft.com/content/82cccbde-b39a-4264-8209-c7d9409eb7ce

The Rodney and Otamatea Times. (1912, August 14). *Coal consumption affecting climate*. Retrieved August 21, 2025, from https://paperspast.natlib.govt.nz/newspapers/rodney-and-otamatea-times-waitemata-and-kaipara-gazette/1912/08/14/7

Thomsen, C. (2013). Sustainability (World Commission on Environment and Development Definition). In: S.O. Idowu, N. Capaldi, L. Zu & A. D. Gupta (Eds.), *Encyclopedia of corporate social responsibility*. https://doi.org/10.1007/978-3-642-28036-8_531

Tveen, L. (2021, August 3). *Green transition in Asia needs a 'buildings first' approach*. The World Economic Forum. Retrieved March 4, 2024, from https://www.weforum.org/agenda/2021/08/green-transition-asia-buildings-first-approach/

UN DESPA. (2018). *The World's Cities in 2018*. The United Nations. Retrieved March 4, 2024, from https://www.un.org/development/desa/pd/sites/www.un.org.development.desa.pd/files/files/documents/2020/Jan/un_2018_worldcities_databooklet.pdf

UN Habitat. (2010, March 18). *Bridging the urban divide: right to the city*. UN Habitat. Retrieved March 4, 2024, from https://staging.unhabitat.org/documents/SOWC10/R12.pdf

UNEP. (2020). *2020 Global status report for buildings and construction* [Towards a zero-emission, efficient and resilient buildings and construction sector]. Retrieved March 1, 2024, from https://globalabc.org/sites/default/files/inline-files/2020%20Buildings%20GSR_FULL%20REPORT.pdf

UNEP, & Yale Center for Ecosystems + Architecture. (2023). *Building materials and the climate: Constructing a new future*. Retrieved August 24, 2025, from https://wedocs.unep.org/20.500.11822/43293.

UNESCO. (n.d.). The built environment - Disaster risk reduction. UNESCO. Retrieved March 1, 2024, from https://en.unesco.org/disaster-risk-reduction/built-environment

USAID. (n.d.). *Energy performance of buildings in Southeast Asia | Basic Page | U.S. Agency for International Development*. USAID. Retrieved March 4, 2024, from https://www.usaid.gov/energy/efficiency/examples/southeast-asia

WGBC. (n.d.). *What is a sustainable built environment?* World Green Building Council. Retrieved March 6, 2024, from https://worldgbc.org/what-is-a-sustainable-built-environment/

World Commission on Environment and Development. (1987). *Our Common Future*. 1987: Brundtland Report. Retrieved April 6, 2022, from https://www.are.admin.ch/are/en/home/media/publications/sustainable-development/rundtland-report.html

Zhao, P., & Zhang, M. (2018, June). The impact of urbanisation on energy consumption: A 30-year review in China. *Urban Climate*, *24*. https://doi.org/10.1016/j.uclim.2017.11.005

Zhou, W., Lee, H. S., & Tian, S. (n.d.). *Accelerating net zero transition of the buildings and construction sector in developing Asia and the Pacific - ADO 2023 Thematic*. Asian Development Bank. Retrieved March 1, 2024, from https://www.adb.org/sites/default/files/institutional-document/874256/adotr2023bp-net-zero-construction-asia-pacific.pdf

CHAPTER 4

Why the Built Environment Matters

"The city throughout the history of mankind has been the meeting place for people. Much of the culture of mankind has happened in the public space" Jan Gehl.

The First Settlements

Early human ancestors, Neanderthals and Denisovans, lived around 2.5 million years ago and the people we call homo sapiens evolved between 200,000 and 300,000 years ago, ironically during a period of climate change (Handwerk, 2021). Since then we have ourselves come up with multiple ways to push ourselves towards extinction. If we avoid nuclear Armageddon we still have anthropogenic climate change to contend with (Wilford, 2002). By contrast, dinosaurs roamed Earth for around 165 million years (US Geological Survey, n.d.) and our current prosperity is dependent on energy derived from the organisms that existed alongside them. Humanity may consider ourselves superior to dinosaurs on many metrics, but to occupy this planet for longer than they did, we need to be better stewards of our environment.

One thing humanity excels at is building. Some structures are built to stand the test of time better than others. Stonehenge is a man-made stone circle in the UK dating from around 5000 years ago, surviving from

S. Crispin, *Sustainability in the Built Environment*,
https://doi.org/10.1007/978-981-95-6549-8_4

the Neolithic to the Digital Age. Older structural remains, pits, in the immediate vicinity date from 8500 to 7000 BC. To this day it is a site for Druid celebrations, yet the structure itself predates Druidism (Wills, 2023). Over millennia it has been repurposed. It has served as a burial site, a place for community meetings and gatherings and a religious site. Currently education and tourism provide a purpose and a revenue stream. In the digital age, signs at the site encourage visitors to take interesting photos to share in cyberspace (Fig. 4.1).

The iconic design of Stonehenge with distinctive horizontal lintels is an early example of sustainability through the ages. Apart from solar orientation, the stones represent complex supply chains, circularity and advanced engineering in a way that sets it apart from the 1000s of other stone circles that remain elsewhere in Europe. The stones are not all locally sourced, 80 huge bluestones were transported from south west Wales, around 150 miles away, a journey that would test modern transportation. The sarsen stones are locally sourced from around 20 miles to the north (English

Fig. 4.1 A leap through 5 Millenia at Stonehenge

Heritage, n.d.). This and other archaeological evidence suggest the bluestones may have been used in another stone circle elsewhere before being moved and reworked by hand in the current site. Mounting the horizontal lintels on the vertical columns, some with mortice and tenon joints, would have taken planning, project management, skilled stone masonry, collaboration and innovative engineering. To this day, site governance of Stonehenge is a source of controversy. There are still calls for an enquiry into police action taken to prevent New Age travellers joining a festival at the site 40 years ago persist (Jewers, 2025). The 1985 "Battle of the Beanfield" demonstrates the tension between the community and the authorities in the use and protection of such assets. These are some of the topics that will be covered in later chapters.

The lessons of human history are not in our favour. Historic climate change is linked to the fall of civilisations. Archaeologists digging into ancient sites have found clues that reveal the now arid places that once had water resources to support ancient cities. The fall of the Maya and Khmer civilisations are linked to drought and a lack of adaptation, Dr Jason Ur of Harvard University is quoted as follows "when we excavate the remains of past civilizations, we rarely find any evidence that they made any attempts to adapt in the face of a changing climate. I view this inflexibility as the real reason for collapse" (Sohn, 2014). Viking settlements in Greenland existed for 400 years during the Medieval Warm Period until around 1250 when the Little Ice Age made it impossible for them to continue to eke out an existence (Greenspan, 2023).

More recently, humanity has excelled at population growth. In 1798 Thomas Malthus, an economist and philosopher, published "An Essay on the Principle of Population". He argued that food security, or the lack of it, would limit population growth and that famine, war and disease would ultimately reverse it (Kagan, 2024). In the end, innovations in healthcare and food production avoided the worst-case scenarios. Cities are the main embodiment of the built environment where they accommodate, feed and improve health outcomes for the growing population. Meanwhile, climate science suggests we are heading to irreversible anthropogenic climate change and calamity, including impacts on food security in some of the most densely populated cities in the poorest parts of the planet. Mass starvation and desperate migration will surely follow. History suggests this would be the end of modern civilisation as we know it.

Cities and Public Health

Much of the success in managing population growth and improving life expectancy is due to the efficiency of cities in managing housing, employment, food distribution and healthcare. The earliest city living communities can be traced to Uruk (modern day Warka, Iraq) where archaeologists have found traces of religious rituals and sacrifice on the site of the White Temple, dated c. 3517–3358 B.C.E. (German, 2015). The role of public health evolved with communal habitation and early religion. Shamanism is the ancient healing practice of indigenous cultures. Modern Shamanism includes connection with nature, what we might call tree-hugging or forest bathing, self-healing, spiritual practice, pilgrimage and ceremony (Dance of the Deer Foundation, n.d.). So popular these days that Shamanism was recently described by The Economist in February 2023 as "Britain's fastest growing religion". The town of Glastonbury in the south west of the UK has witnessed the interweaving of Celtic myth, legend, historic and modern Christianity and spirituality over centuries. The Glastonbury Trust is dedicated to promoting "emotional well-being, citizenship, environmental improvement and spiritual growth" (The Glastonbury Trust, n.d.). The Trust owns a courtyard of retail properties, The Glastonbury Experience, that houses a range of spiritual and healing centres offering courses in contemporary Shamanism.

Public Health evolved through the ages in urban centres from basic sanitation, ancient building codes and water supplies, to early Islamic medicine and monastery hospitals that led to isolation for sufferers of leprosy, for example, until the emergence of modern western medicine (Tulchinsky & Varavikova, 2014). A formal and long-standing relationship between public health and city development can also be seen in ancient Greek and Roman Civilisations. Roman sanitary engineering is well known as is the development of hospitals and medical practice in both Islamic and Christian cultures (Tulchinsky & Varavikova, 2014).

Cities grow at the expense of natural habitats and are also a breeding ground for disease. Globally, around 75 million people are thought to have died during The Black Death of 1347 to 1352, with China losing half and Europe losing one-third of their respective populations (Tulchinsky & Varavikova, 2014). Apart from the Black Death, Cholera outbreaks focused on cities. In London, British physician John Snow in 1854 mapped Cholera clusters resulting in the identification of "a public water pump as the source of the outbreak" (Kochtitzky et al., 2006).

Spanish Flu and most recently the COVID-19 pandemic reminded us again of the link between healthy cities and healthy people (Clark, 2022) and is just the latest in a long series of public health crises that have shaped urban centres. COVID-19 responses include improving food security, increasing local food production in cities as diverse as Singapore and Lagos and increasing spaces for active transport in the previously motor vehicle dominant streets of major cities worldwide. Low Traffic Neighbourhoods in London established in the pandemic remain a source of community tension.

What is the link to sustainability and the built environment? An October 2023 editorial in the British Medical Journal called for the climate and nature crises to be considered a single global health emergency rather than treating them separately. Taking a systems approach and treating climate and biodiversity as a complex interdependent system can help find solutions that unlock the regenerating potential of nature (Abbasi et al., 2023, 2355).

At the same time, continuing development of built environments threatens to diminish the health benefits of high-quality green spaces that help filter the air, reduce heat islands and provide space for physical activity (Macguire et al., 2022). Connectivity to nature promotes social interaction and reduces stress, loneliness and depression (Wong et al., 2018). A study in the Lancet of 93 European cities with an average tree cover of 15% revealed that increasing tree cover to 30% would lower average temperatures by 0.4% which could reduce heat-related deaths by around 35%. Cities in southern Europe tend to experience temperatures that are around 1.5 °C higher than the surrounding countryside. Cities with lower tree cover suffer from more extreme heat islands. Both London and Barcelona are among cities that are aiming for 30% tree cover as they experience more intense summer heatwaves. Green roofs are another means of mitigating urban heat islands.

Stephen White is an Architect and Founder and former Managing Director of IACTION, (Ideas in Action) and lived in China from 1990 to 2020. He witnessed the greening of Shanghai when legislation was introduced to require parks within walking distance of new residential areas and the planting of trees along new and widened streets. When air quality was monitored and data published, awareness of "bad air days" increased and indoor air quality became a concern for tenants and their staff (Fig. 4.2).

Fig. 4.2 View of Lujiazui, Shanghai, China (authors own photograph)

In recognition of the connection between the climate crisis and health, the COP28 UN Climate Conference held its first ever "Day of Health" with a number of ways that health is impacted by climate:

1. Air quality.
2. Heatwaves.
3. Infectious Diseases.
4. Food and water.
5. Mental health (World Economic Forum, 2023).

All of these health impacts are worse in the built environment; the built environment is also key to mitigating the health impacts of climate change.

Urban Pollution and the Planning Response

Cities are bad for our health. Environmental problems associated with the urban environment are broad ranging. Air pollution resulted from the industrial revolution and rapid urbanisation across Europe from the late 1700's as farm workers moved to cities to take up dirty and dangerous industrial work that used coal. This became a public health issue. Pollution in London got so bad that the death rate in December 1873 was 40% above normal, similar incidents were recorded in 1880, 1882, 1891, 1892 and 1948 (Met Office, n.d.). A further smog in 1952 "The Great Smog" showed that serious smog had become a lethal danger to public health and cost to society (Martinez, 2024) and led to Government action in the form of the Clean Air Acts in the UK (1956, 1968 and 1993) (Air-Quality, n.d.) and US Air Pollution Act (1955) which is said to have "saved millions of lives and trillions of dollars" (Wells et al., 2020). The smog cloud has moved to major Asian cities such as Beijing and Delhi in the more recent past. By the 1950s the oil revolution had taken over from coal and mass private vehicle ownership was taking off. Since then coal fired power generation has tapered off in Europe to be replaced with natural gas, nuclear and, most recently, renewables.

Expanding built environments resulted in social ills and tragic health outcomes. In London, up to 75% of children died before the age of 5 in 1723 (Difford, n.d.). These were documented by William Hogarth's A Harlot's Progress in 1733 which inspired the later George Cruikshank's "The Bottle" published in 1847 (British Library, n.d.). Similarly in the US, Nathaniel Currier's 1846 Drunkard's Progress illustrated the decline of a man from the first glass of wine to eventual suicide. As with environmental issues, social problems were resolved with various governance responses. These included the UK Gin Acts of 1729, 1736, 1743 and 1751 which aimed to combat the death rate of babies and infants. Another urban response was the later establishment of the first professional modern police force in 1829 by Sir Robert Peel as a means to enforce the increasing number of laws needed to govern dense urban populations (Greater Manchester Police Museum & Archives, n.d.).

Early urban planning shows socially responsive approaches that supported a business model. One resulted in the village of Port Sunlight, founded by William Hesketh Lever in 1888 to provide decent and affordable housing for Lever Brothers soap factory workers, an early form of "prosperity sharing" as Lever himself described it (Port Sunlight Village Trust, n.d.). Similarly, the village of Bournville, founded in 1893 by the Quaker Cadbury family to support a chocolate factory that had moved out of the slums of Birmingham in 1879 (Robinson, 2010). Bournville supported a diverse community, beyond just factory employees, through low housing rents and planning provided for parks and recreation areas to support what we would now call "active lifestyles" with sport, swimming and walking encouraged (Bournville Village Trust, n.d.). What both these, and other, cases, is a planned business response to linked environmental and social issues that resulted in improved outcomes for both business and community.

Fast forward to the present day, and the modern equivalent of 1950's London "pea soupers" has spread worldwide. WHO statistics suggest that 7 million people a year die from fine particulate matter (PM 2.5 is the most dangerous) in polluted air and 91% of the world's population live in places where pollution exceeds WHO air quality guidelines. According to IQ Air, an air technology company who sell air cleaning products and monitor air quality, only 7 countries met WHO guidelines in 2023 and the 3 most polluted countries were Bangladesh, Pakistan and India (Montenegro, 2024). 99 out of 100 of the most polluted cities were in Asia, 83 of which were in India and in South-east Asia. Indonesia, Vietnam and Thailand all cities have PM2.5 pollution that is 10 times worse than WHO guidelines (Regan, 2024). Sources of pollution are burning fossil fuels for electricity generation, burning agricultural waste, wildfires and indoor cooking with dirty fuel sources which can lead to higher indoor pollution than that outdoors.

My own experience living in mainland China from 1994 to 2017 was that observable air pollution increased with development of the built environment and related economic activity. Concerted government action was required and in particular inter-provincial regulation to improve air quality from what had been "bad air days" also known as the "airpocalypse" (FlorCruz, 2013). China managed to reduce air pollution by 42% between 2013 and 2021 resulting in an increase in life expectancy of 2.2 years (Yeung, 2023).

China still faces many challenges, having added six times as much coal fired electricity generation in 2022 than the rest of the world combined. China accounts for 55% of global coal consumption and most mainland Chinese cities do not meet the Chinese national standard for PM2.5. Under the Paris Agreement, China will have to set 2035 climate targets but the clear tension remains between post-pandemic economic recovery (job creation) and meeting goals to reduce energy intensity (energy consumed per unit of GDP) (Shi, 2024) (Fig. 4.3).

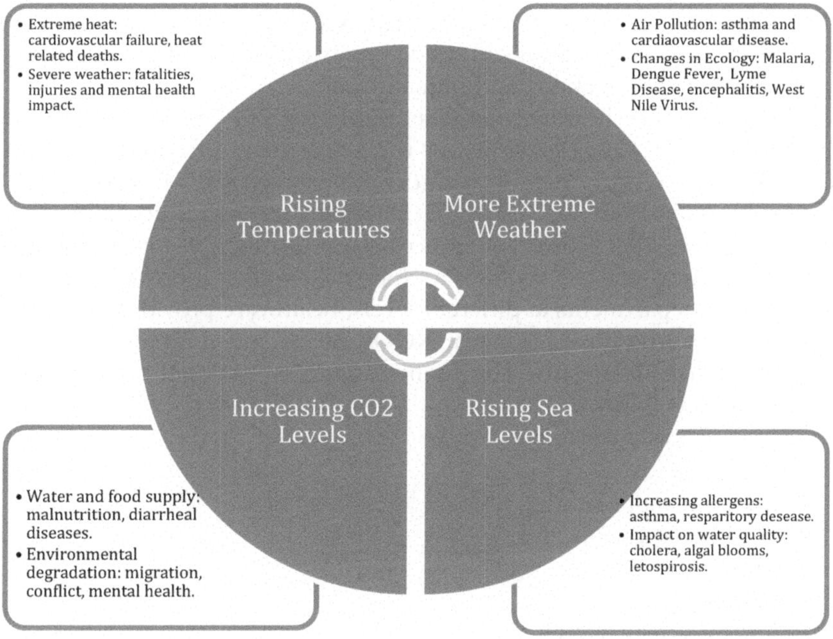

Fig. 4.3 Impact of climate change on human health (Centers for Disease Control and Prevention, n.d.)

Healthy Cities

The quality of the built environment is closely related to public health and the integration of city and public health is frequently referred to as "healthy cities". The WHO describes a healthy city as a process rather than a health standard. It is a city that seeks to improve the health of residents by continually creating and improving physical and social environments, delivering for both people and the planet. As well as mobilising local resources to help residents maximise their potential in their life functions. This includes people, participation, prosperity, planet, place and peace. Under planet, "a healthy city ensures that the health and well-being of both the people and the planet are at the heart of all the city's internal and external policies. This includes, a whole-of-city approach to health and well-being; coherence across levels of governance in the approach to health and well-being and strengthened city healthy diplomacy" (WHO, n.d.). Environmental conditions are listed as an indicator of a healthy city thus providing a clear link, if still needed, between the built environment, environmental sustainability and human health.

The approach to healthy cities is set out in the 2007 second edition of Building Healthy Cities published by the Department of Health in Hong Kong after the Severe Acute Respiratory Syndrome (SARS) in 2003. The document states "any city can be a Healthy City if it is committed to health and has a structure and process to work for its improvement". Putting this into practice requires a multi-stakeholder approach between government departments at different levels, healthcare providers, occupational health authorities, academia, NGOs, media and community groups including religious groups. The Hong Kong case sets out three phases and 20 steps as follows:

Phase 1 Getting started.

1. Building support.
2. Healthy cities concepts.
3. Know your city.
4. Funding.
5. Project "ownership".
6. Project proposal.
7. Project approval.

Phase 2 Getting organised.

8. Selecting a steering committee.
9. Project environment.
10. Work definition.
11. Project office.
12. Project strategy.
13. Building project capacity.
14. Accountability mechanisms.

Phase 3 Taking action.

15. Raising awareness.
16. Strategic planning.
17. Cross sectoral action.

18. Community participation.
19. Innovation.
20. Healthy public policy (Department of Health, HK SAR, 2007).

The clustering of humans in communities, filth, disease and health, public health approaches, industrial pollution, urban planning and, in the next section, anthropogenic climate change are all part of the same cycle.

Historic and Future Urbanisation Trends with Examples from Rapidly Urbanising Countries in Asia and Africa

The imperative for Asia to decarbonise is set out in Decarbonising Asia, edited by Tony A Very and Roman Y Shemakov: "As more than half of the planet's CO2 is emitted in Asia–Pacific, there is no way around it: the region has to become a leader in the global fight against climate change, if we are to win it. This transition also presents one of the greatest investment opportunities of the coming decades. The region is the most vulnerable to desertification, flooding, and sea level rise, but also has most of the planet's people, megacities, and renewable energy capacity. The sky is the limit".

In 2023 nearly 56% of the world's population lived in cities and expected to double by 2050, already 80% of GDP is city based (The World Bank, 2023). UN-Habitat describes urbanisation in Asia Pacific as a megatrend with an ever-growing urban population. The following paragraphs highlight some of the challenges and approaches from Bangladesh, China, India and Pakistan.

Bangladesh (population 163 million) is especially exposed to increased frequency of flooding resulting from climate change. Urban population rates are projected to reach 60% by 2050 compared to 36% in 2024 (Islam, 2024). Four Bangladeshi cities have their own development authorities and tailored master plans (Hasan & Monowar, 2022). One of these, Rajshahi, has a City Resilience Strategy, dated 2016, prepared in partnership by Local Governments for Sustainability (ICELEI), Asian Cities Climate Change Resilience Network and Rockefeller Foundation. The study identified extreme risks in 4 out of 5 urban systems (water, health, biodiversity and economy). The 5th is solid waste management which is high risk. There are 22 resilience interventions to help the city adapt summarised under the following headings:

1. Rainwater harvesting and water course protection.
2. Health response preparedness and risk management.
3. Solid waste management and awareness.
4. Biodiversity master plan, communication and awareness.
5. Develop climate adaptive business, e.g. eco-tourism (ICLEI, 2016).

Identifying the risks and the mitigation is just one step, the key is getting the mitigation integrated into the city planning for implementation.

Given the size of population, China is a key driver of global urbanisation and Chinese urbanisation rates reaching nearly 65% in 2022 (The State Council, 2023) compared to just over 10% in 1949. The concept of sustainable urban development in China is not new, Cai and Tang (2021) identify the phases of development, eco-cities (1990s), low carbon cities (2000s) and smart eco-cities (2010s). The question has tended to be what exactly does it mean?

The following timeline, summarised from Ren and Kim (2023), follows China's Five-Year Plan approach and might help the reader understand the efforts the Chinese Government have set in place:

2001 to 2005 10th Five-Year Plan sets out the importance of green buildings and the policy framework. Emphasis on energy saving, technology and financial support. Involvement of academia and others in the concept and theory of green buildings.
2006 to 2010 awareness and promotion of green building development, supporting policies and technology development. Continued focus on energy efficiency, technology and sustainable design.
2011 to 2015 increasing focus on green building standards, energy saving and insulation materials. The "Comprehensive Building Evaluation Criteria" and "Building Energy Efficiency Evaluation Criteria" were updated in 2014. Model buildings were showcased at the Beijing Olympic Village and the Shanghai World Expo.
2016 to 2020 continued research on green buildings, updated Green Building Evaluation Criteria and report published on building energy efficiency development status and challenges. Building materials were considered in depth and an increasingly systematic approach to green buildings.

2021 to 2022 higher standards for green buildings, integration of energy saving into design, technology solutions for management and emphasis on sustainable architecture. Further emphasis on building operations (Ren & Kim, 2023).

The Chinese government-led top-down approach should be viewed positively representing a firm commitment to sustainable development at a time of rapid urbanisation, as such it provides a model for others, but it lacks local contextual consideration. It uses a simple indicator to all the vast and varied geographies of China with no nuance for local needs. Local governments have been known to use the eco-city and low carbon labels as a means to promote large-scale new developments with little local demand resulting in increasing local government debt (Ren & Kim, 2023).

Few countries have the scale of pressure on urban land, infrastructure and services as India. Urbanisation in India lags only that of China with 530 million city dwellers compared to 942 million in China (Statistics Times, 2024). In 2023 just over 36% of the Indian population were urbanised (O'Neill, 2025). The World Bank forecasts this to rise to 40% by 2036 and also the urgent need to build new infrastructure as a key for "liveable, climate resilient and inclusive cities" including water resources, efficient public transport, healthcare and work skills (Kouame, 2024). Responses have included updating city governance to give more oversight at the municipal level and a more inclusive approach to non-government organisations, including working with communities, universities, private sector and research institutes (Aijaz, 2025).

At the central government level, the ClimateSmart Cities Assessment Framework, developed by the Ministry of Housing and Urban Affairs in association with the Climate Centre for Cities and the National Institute of Urban Affairs, provides a roadmap for Indian cities to combat climate change. The framework includes assessment tools based on planning, green cover and biodiversity, energy and green buildings, air quality and mobility, water and waste management. The assessment includes mitigation and adaptation to enable cities to progressively improve their scores (MoHUA, 2022).

Notable Indian city cases include Nagpur which has set a strategy to achieve city-wide net zero buildings by 2050 (Bose, 2024). The plan identifies a potential annual GHG mitigation potential of 30% from residential buildings, 10% from commercial buildings and a huge 44% from solid waste (Urban LEDS, n.d.).

The plan for residential buildings includes:

1. Optimise energy efficiency.
2. Rooftop solar.
3. Implement Guidelines for Climate Responsive Homes in Nagpur.
4. Cool rooves.
5. Affordable housing to implement green building principals.

And for commercial buildings:

6. District cooling systems.
7. Adopt energy efficient plant and machinery.
8. Benchmark building energy consumption.
9. Scale up rooftop solar.
10. Encourage use of evaporative coolers (Urban LEDS, n.d.).

An observer might note that the above 10 points could be considered "low-hanging fruit" but the economics are contextual. Installing a cool roof on different buildings costs the same regardless of any difference in value. Investment by commercial building owners is needed along with a regulatory push and an incentive pull.

Pakistan was hit hard by flooding in 2022 with a heavy monsoon and melting glaciers that covered a third of the country affecting 33 million people (UNICEF, 2023) and estimated damage of US$40 billion. At a city level the Karachi Climate Action Plan is a 152-page document approved in October 2023 that sets out the ambition to reduce greenhouse gas emissions by 3.7% by 2030, "34.8% by 2040 and 62.2% by 2050". It is an impressive and comprehensive document that uses the C40 Cities Climate Action Planning Framework (The Urban Unit, Government of Punjab, 2025). It includes regulatory amendments, governance changes, goals and KPI's for monitoring progress. The extensive Green Building guidelines for Pakistan are covered in Chapter 8.

The message from Africa is little different. Nigeria's population is forecast to grow from 216 million in 2023 to 400 million in 2050 (UNFPA,

2023). Between 1990 and 2022 the number of African cities doubled and the continents urban population tripled (OECD, 2022).

The economic benefits of urbanisation in Africa cannot be disputed. According to the report Africa's Urbanisation Dynamics 2022: The Economic Power of Africa's Cities, the benefits include:

1. Hourly wages in large cities are more than twice those of rural areas.
2. Young city dwellers receive up to 4 years more education that their rural counterparts.
3. 80% of households in large cities are connected to the grid compared to 20% in rural areas.
4. Increasing clustering of cities enables industry specialisation and economic integration so smaller cities can deliver economic benefits too. Free Trade Agreements result in cross-border economic benefits.
5. Smaller educational gender gap in cities.
6. Product innovation is almost twice as high in cities over 1 million inhabitants compared to those with less than 250,000 inhabitants (OECD, 2022).

The above is just a sample of benefits. Let's balance that with the sustainability challenges.

On the economic side, income polarisation within cities results in poverty and slums (Odunbaku & Kiwala, 2015). Kamana et al, (2024) used a Systematic Literature Review to analyse 243 urban study papers published between 2000 and 2023 covering 107 African cities to summarise the common sustainability challenges affecting cities across the continent. The same methodology was used to identify urban strategies. The various challenges were clustered under headings of Environment—Climate Change (ECC), Politico-Socio Economic (PSE) and Spatial Planning—Governance (SPG). The results are ranked as follows:

1. Environment Degradation/Pollution (ECC)
2. Slums/Informal settlements (PSE)
3. Drought/water scarcity (ECC)
4. Rapid urbanisation/sprawl (ECC)
5. Socio-spatial inequality (PSE)
6. Rapid population growth/inward migration (SPG)

7. Food insecurity (PSE)
8. Urban flooding/landslides (ECC)
9. Traffic congestion (SPG)
10. Waste management (SPG)
11. Inadequate infrastructure (SPG)
12. Crime (SPG)
13. Agricultural land grabbing (ECC)
14. Unaffordable housing (PSE)
15. City centre decline (SPG)
16. Evictions (PSE)
17. Unemployment/poverty (PSE) (Kamana et al, 2024)

Many of these challenges are interlinked and, left unmanaged, may be part of a vicious cycle. Such local challenges require local solutions that reflect the contextual nature of each city and its specific problems. The most common strategies in African cities were found to be inner city redevelopment and setting green, sustainable or resilient city goals.

It's easy to focus too much on the challenges, but the message is one of innovation, solutions and finding optimism everywhere. Just as early built environment social and environmental ills were dealt with, so today we see hope even in some unlikely places. The urban population of sub-Saharan Africa is expected to grow by 270 million between 2030 and 2035 (R, 2023), but there is plenty of evidence that the challenges thrown up by buildings and the urban environment can be managed.

1. Nairobi energy is 70% renewable.
2. Cairo enjoys comprehensive public transport and per capita energy consumption that is nearly half the global average.
3. Green City Kigali occupies 600 hectares and aims to be carbon neutral by 2050.
4. Cocody in Ivory Coast plans to cut carbon emissions by 70% by 2030.
5. Dakar is aiming to cut diesel power generation from 90% in 2013 to 5% by 2035.
6. Kampala has an energy and climate change action plan, first published in 2016 (Uganda as a whole uses over 90% hydropower).
7. Tsevie in Togo uses biomass for 64% of energy consumption (Marchant, 2021) and (R, 2023).

At a country level, it should be noted that Rwanda is nearly 35% urbanised already and is using biogas, geothermal, solar and hydropower to achieve nearly 50% renewable energy supply by 2024 (R, 2023).

As a model of sustainability and smart city innovation, Konza Technolopis, Kenya, is a model for balancing growth and the environment with the following features:

1. Walking or cycling access to all essential services.
2. Walkable, connected environment.
3. Integration of city management, economic growth and smart infrastructure.
4. Leading technology, innovation and talent development.
5. Green spaces and buildings.
6. Advanced waste and e-waste management (Umoren-Udo, 2025).

The Role of the Built Environment in Carbon and Greenhouse Gas Emissions

As described in Chapter 3, the scale of urbanisation is increasing to accommodate a growing population. In 1987, at the time of the Brundtland report, the world population was around 5 billion, grew to 7.7 billion in 2020 and is expected to reach 9.7 billion by 2050 (UN, n.d.). Of whom around 4.2 billion lived in cities in 2018 with an additional 2.5 billion urban residents by 2050 (UN, 2018).

While cities generate vast wealth, Indian cities are expected to generate 70% of GDP by 2030 (Sankhe et al., 2010), they are also voracious in their use of energy. 70% of global CO2 emissions are generated by cities (Net Zero Cities, n.d.). Globally, 40–50% of energy used is building related and in densely populated Asian cities like Hong Kong and Singapore, this rises to around 90%. Energy consumption results in greenhouse gas emissions measured carbon equivalent (CO2e). Building-related carbon emissions can be divided into embodied carbon, related to the building materials and construction processes and operational carbon from energy consumption in the day to day use of the building (RICS, 2017).

The International Energy Agency (IEA) was established in response to the 1974 oil crisis to co-ordinate a response to oil security concerns. It now considers all fuels including renewables and issues like demand management. Buildings are one of the sectors the IEA monitors and reports that building operations account for 30% of global energy consumption and 26% of global energy-related emissions. It also tracks progress towards the targets for "all new buildings and 20% of existing buildings to be zero-carbon ready by 2030". Zero-carbon ready means supplied by renewable energy or all electric with the potential to decarbonise. In terms of CO2 direct emissions from building operations, the period 2015 to 2021 saw average increases of 1% per year and a small decrease in 2022 (Delmastro & Chen, 2023).

Summary

This chapter has linked the emergence of Homo Sapiens, human clustering in cities resulting in social ills and health crises to historic and modern urban planning responses which cover food security and healthy cities in Africa and Asia. It is not difficult to think the scale of the problem is too vast to handle, 70% of greenhouse gas emissions are generated in cities (The World Bank, 2023). Yet there are plenty of cases of manageable solutions in unexpected places as described above and as will be discussed in later chapters.

All the studies and plans needed to understand the problems and way forward have been completed, written up and published, most of which is freely available. All the supporting evidence exists and this surely points to the sustainability solutions that can be adopted with national policy and city level implementation frameworks. Cutting built environment related carbon emissions by 50% has the potential to reduce global emissions by 16 to 19%. Yet so many remain sceptical or think it is someone else's problem. This is why it is important, not just because of the scale of problem and the potential to reduce emissions, but because of the requirement for everyone to do something to implement the available solutions.

References

Abbasi, A. K., Barbour, V., Benfield, T., Bibbins-Domingo, K., & Hancocks, S. (2023, October 25). Time to treat the climate and nature crisis as one indivisible global health emergency. *British Medical Journal*, 2355. https://doi.org/10.1136/bmj.p2355

Aijaz, R. (2025, June 17). *Fixing the foundations: Reforming urban governance in India*. Observer Research Foundation. Retrieved June 21, 2025, from https://www.orfonline.org/expert-speak/fixing-the-foundations-reforming-urban-governance-in-india

Air-Quality. (n.d.). *Air quality | History of air pollution in the UK*. Air Pollution. Retrieved May 27, 2024, from https://www.air-quality.org.uk/02.php

Bose, M. (2024, March 4). *Nagpur takes lead with India's first city-specific Zero Carbon plan for buildings*. Deccan Herald. Retrieved June 21, 2025, from https://www.deccanherald.com/india/maharashtra/nagpur-takes-lead-with-indias-first-city-specific-zero-carbon-plan-for-buildings-2920736

Bournville Village Trust. (n.d.). *The Bournville Story*. Bournville Village Trust. Retrieved September 20, 2022, from https://www.bvt.org.uk/our-business/the-bournville-story/

Cai, Z., & Tang, Y. (2021, May). Toward a sustainable city: A scoping review of eco-cities development and practices in China. *Chinese Urban Planning and Construction*. https://doi.org/10.1007/978-3-030-65562-4_9

Clark, G. (2022, October 7). *The century of cities - Connected Places Catapult*. Connected Places Catapult. Retrieved October 17, 2022, from https://cp.catapult.org.uk/article/the-century-of-cities/

Dance of the Deer Foundation. (n.d.). *SHAMANISM | What is Shamanism?* shamanism. Retrieved May 27, 2024, from https://www.shamanism.com/what-is-shamanism

Delmastro, C., & Chen, O. (2023, July 11). *Buildings - Energy system - IEA*. International Energy Agency. Retrieved March 20, 2024, from https://www.iea.org/energy-system/buildings

Department of Health, HK SAR. (2007). *Building healthy cities guidelines*. Centre for Health Protection. Retrieved May 27, 2024, from https://www.chp.gov.hk/files/pdf/building_healthy_cities_guidelines.pdf

Difford, S. (n.d.). *History of gin (1728 - 1794) - London's gin craze*. Difford's Guide. Retrieved September 14, 2022, from https://www.diffordsguide.com/g/1108/gin/history-of-gin-1728-1794

English Heritage. (n.d.). *The stones of Stonehenge*. English Heritage. Retrieved September 17, 2025, from https://www.english-heritage.org.uk/visit/places/stonehenge/things-to-do/stone-circle/stones-of-stonehenge/

FlorCruz, J. A. (2013, January 19). *Living with Beijing's 'air-pocalypse'*. CNN. Retrieved May 27, 2024, from https://edition.cnn.com/2013/01/19/world/asia/china-florcruz-pollution/index.html

German, S. (2015, August 8). *Smarthistory – White Temple and ziggurat, Uruk*. Smarthistory. Retrieved May 27, 2024, from https://smarthistory.org/white-temple-and-ziggurat-uruk/

Greater Manchester Police Museum & Archives. (n.d.). *The Peeler | Greater Manchester police*. Greater Manchester Police Museum. Retrieved September 14, 2022, from https://gmpmuseum.co.uk/collection-item/the-peeler/

Greenspan, J. (2023, April 19). *Why Did the Vikings Disappear From Greenland?* History.com. Retrieved June 20, 2025, from https://www.history.com/articles/why-did-the-viking-disappear-from-greenland

Handwerk, B. (2021, February 2). *An evolutionary timeline of Homo Sapiens | Smithsonian*. Smithsonian Magazine. Retrieved May 27, 2024, from https://www.smithsonianmag.com/science-nature/essential-timeline-understanding-evolution-homo-sapiens-180976807/

Hasan, M. R., & Monowar, N. (2022, January 27). *Urban planning in Bangladesh: Challenges and opportunities*. The Business Standard. Retrieved June 20, 2025, from https://www.tbsnews.net/supplement/urban-planning-bangladesh-challenges-and-opportunities-362911

ICLEI. (2016, November). *City resilience city resilience strategy*. ICLEI South Asia. Retrieved June 20, 2025, from https://southasia.iclei.org/wp-content/uploads/2021/12/City-Resilience-Strategy_Rajshahi.pdf

Islam, M. T. (2024, June 6). *Cities in Bangladesh must refocus to combat climate change | Epicenter*. Epicenter. Retrieved June 20, 2025, from https://epicenter.wcfia.harvard.edu/blog/cities-bangladesh-must-refocus-combat-climate-change

Jewers, N. (2025, June 2). *Battle of the Beanfield: Calls for a public inquiry into violence*. BBC. Retrieved September 17, 2025, from https://www.bbc.co.uk/news/articles/c780ve9nde0o

Kagan, J. (2024, April 16). *Who is Thomas Malthus? What is the Malthusian growth model?* Investopedia. Retrieved May 27, 2024, from https://www.investopedia.com/terms/t/thomas-malthus.asp

Kamana, A. A., Radoine, H., & Nyasulu, C. (2024). Urban challenges and strategies in African cities – A systematic literature review. *City and Environment Interactions, 21*. https://doi.org/10.1016/j.cacint.2023.100132

Kochtitzky, C., Frumkin, H., & Rodriguez, R. (2006, December). Urban planning and public health at CDC. *MMWR Supplements*, *55*(2), 34–38. 17183243

Kouame, A. (2024, January 30). *Gearing up for India's rapid urban transformation*. World Bank. Retrieved June 21, 2025, from https://www.worldbank.org/en/news/opinion/2024/01/30/gearing-up-for-india-s-rapid-urban-transformation

Macguire, F., Mulcahy, E., & Rossington, B. (2022). The Lancet countdown on health and climate change. Policy Brief for the UK 2022. *The Lancet*. https://s41874.pcdn.co/wp-content/uploads/Lancet-Countdown-2022-UK-Policy-Brief_EN.pdf

Marchant, N. (2021, April 27). *6 African cities leading the way to a green future | World Economic Forum*. The World Economic Forum. Retrieved May 27, 2024, from https://www.weforum.org/agenda/2021/04/africa-cities-renewables-ren21/

Martinez, J. (2024, April 26). *Great smog of London | 1952, Cause, Deaths, & Facts*. Britannica. Retrieved May 27, 2024, from https://www.britannica.com/event/Great-Smog-of-London

Met Office. (n.d.). *The great smog of 1952*. Met Office. Retrieved September 14, 2022, from https://www.metoffice.gov.uk/weather/learn-about/weather/case-studies/great-smog

MoHUA. (2022). *ClimateSmart cities assessment framework 3.0*. Retrieved June 12, 2025, from https://niua.in/c-cube/sites/all/themes/zap/assets/pdf/CSCAF_3_0_Technical_document.pdf

Montenegro, M. (2024, March 19). *2023 IQAir World Air Quality Report*. IQAir. Retrieved May 27, 2024, from https://www.iqair.com/us/newsroom/waqr-2023-pr

Net Zero Cities. (n.d.). *Towards climate neutral European cities by 2030*. NetZeroCities: Home. Retrieved June 21, 2025, from https://netzerocities.eu/

Odunbaku, O., & Kiwala, L. (2015). *Sustainable urban development in Africa*. UN Habitat. Retrieved May 27, 2024, from https://unhabitat.org/sites/default/files/download-manager-files/Sustainable%20Urban%20Development%20in%20Africa.pdf

OECD. (2022, April 26). *Africa's Urbanisation Dynamics 2022 - The economic power of Africa's cities*. African Development Bank. Retrieved May 27, 2024, from https://www.afdb.org/en/documents/africas-urbanisation-dynamics-2022-economic-power-africas-cities

O'Neill, A. (2025, June 13). *India - Urbanization 2023*. Statista. Retrieved June 21, 2025, from https://www.statista.com/statistics/271312/urbanization-in-india/

Port Sunlight Village Trust. (n.d.). *Welcome and introduction to Port Sunlight*. Port Sunlight. Retrieved September 20, 2022, from https://www.portsunlightvillage.com/about-port-sunlight/

R, M. (2023, December 15). *Urban sustainability: Lessons from Africa's greenest cities*. Capmad.com. Retrieved May 27, 2024, from https://capmad.com/sustainable-development-en/urban-sustainability-lessons-from-africas-greenest-cities/

Regan, H. (2024, March 18). *The world's 100 worst polluted cities are in Asia — And 83 of them are in just one country*. CNN. Retrieved May 27, 2024, from https://edition.cnn.com/2024/03/18/climate/air-pollution-report-2023-asia-climate-intl-hnk/index.html

Ren, W., & Kim, K. (2023, September 9). A study on the green building trend in China—From 2001 to 2022, focusing on research topic words. *Sustainability, 15*. https://doi.org/10.3390/su151813505

RICS. (2017). Whole Life Carbon Assessment for the Built Environment. *RICS professional standards and guidance, UK*. rics.org/guidance

Robinson, J. (2010, January 22). Bournville: The town that chocolate built | Cadbury. *The Guardian*. https://www.theguardian.com/business/2010/jan/23/bournville-cadbury-town

Sankhe, S., Vittal, I., Dobbs, R., Mohan, A., Gulati, A., Ablett, J., Gupta, S., Kim, A., Paul, S., Sanghvi, A., & Sethy, G. (2010, April 1). *India's urban awakening: Building inclusive cities, sustaining economic growth*. McKinsey. Retrieved May 27, 2024, from https://www.mckinsey.com/featured-insights/urbanization/urban-awakening-in-india

Shi, J. (2024, March 19). *Opinion | Why public participation is crucial in China's climate change fight*. South China Morning Post. Retrieved May 27, 2024, from https://www.scmp.com/comment/article/3255922/it-could-be-watershed-year-chinas-climate-change-fight-public-participation-crucial

Sohn, E. (2014, January 20). *Climate change and the rise and fall of civilizations – Climate Change: Vital signs of the Planet*. Climate Change. Retrieved June 20, 2025, from https://climate.nasa.gov/news/1010/climate-change-and-the-rise-and-fall-of-civilizations/

Statistics Times. (2024, February 3). *Countries by urban population 2024*. StatisticsTimes.com. Retrieved June 21, 2025, from https://statisticstimes.com/demographics/countries-by-urban-population.php

The State Council. (2023, February 6). *Urbanization to be major growth driver*. State Council Information Office. Retrieved May 27, 2024, from http://english.scio.gov.cn/chinavoices/2023-02/06/content_85090951.htm

The World Bank. (2023, April). *Urban development overview*. World Bank. Retrieved May 27, 2024, from https://www.worldbank.org/en/topic/urbandevelopment/overview

The Urban Unit, Government of Punjab. (2025). *Climate Action Plan (KCAP)*. United Nations Development Programme. Retrieved June 21, 2025, from https://www.undp.org/sites/g/files/zskgke326/files/2025-05/karachi_climate_action_plan_kcap.pdf

The British Library (n.d). *A Harlot's progress*. The British Library. Retrieved September 14, 2022, from https://www.bl.uk/collection-items/a-harlots-progress

The Glastonbury Trust. (n.d.). *Glastonbury is a place of unlimited potential*. The Glastonbury Trust. Retrieved August 12, 2025, from https://www.glastonburytrust.org/

Tulchinsky, T., & Varavikova, E. (2014, October 10). A history of public health. *The New Public Health*. https://doi.org/10.1016/B978-0-12-415766-8.00001-X

Umoren-Udo, G. (2025, January 31). *Konza Technopolis: Pioneering sustainability and smart city innovation in Kenya*. Retrieved June 15, 2025, from https://www.linkedin.com/pulse/future-cities-konza-kenya-esg-real-estate-laboratory-o9uyf/

UN. (2018). *68% of the world population projected to live in urban areas by 2050, says UN | United Nations*. the United Nations. Retrieved May 27, 2024, from https://www.un.org/uk/desa/68-world-population-projected-live-urban-areas-2050-says-un

UNFPA. (2023, February 3). *United Nations Population Fund Country programme document for Nigeria*. UNFPA Nigeria. Retrieved May 27, 2024, from https://nigeria.unfpa.org/en/publications/united-nations-population-fund-country-programme-document-nigeria

UNICEF. (2023, August 25). *Devastating floods in Pakistan | UNICEF*. Unicef. Retrieved June 21, 2025, from https://www.unicef.org/emergencies/devastating-floods-pakistan-2022

United Nations. (n.d.). *Population | United Nations*. The United Nations. Retrieved May 10, 2022, from https://www.un.org/en/global-issues/population

Urban LEDS. (n.d.). *Climate resilient city action plan – Nagpur*. Urban LEDS. Retrieved June 21, 2025, from https://urban-leds.org/wp-content/uploads/2022/02/Summary-Climate-Resilient-Cities-Action-Plan_Nagpur.pdf

US Geological Survey. (n.d.). *When did dinosaurs become extinct? | U.S. Geological Survey*. USGS.gov. Retrieved October 28, 2022, from https://www.usgs.gov/faqs/when-did-dinosaurs-become-extinct

Wells, J., Gardiner, B., & Archive, B. (2020, December 31). *The Clean Air Act has saved millions of lives and trillions of dollars*. National Geographic. Retrieved May 27, 2024, from https://www.nationalgeographic.com/environment/article/clean-air-act-saved-millions-of-lives-trillions-of-dollars

WHO. (n.d.). *What is a healthy city*. World Health Organization (WHO). Retrieved May 27, 2024, from https://www.who.int/europe/groups/who-european-healthy-cities-network/what-is-a-health-city

Wilford, J. N. (2002, February 26). When humans became human. *The New York Times*. https://www.nytimes.com/2002/02/26/science/when-humans-became-human.html

Wills, M. (2023, October 14). *Stonehenge before the Druids (long, long, before the Druids)*. Retrieved September 17, 2025, from https://daily.jstor.org/stonehenge-before-the-druids-long-long-before-the-druids/

Wong, F., Yuen, J., Chang, K., & Wong, F. (2018, September 12). Assessing quality of life using WHOQOL-BREF: A cross-sectional study on the association between quality of life and neighborhood environmental satisfaction, and the mediating effect of health-related behaviors. *BMC Public Health*. https://doi.org/10.1186/s12889-018-5942-3

World Economic Forum. (2023, November 13). *5 ways climate change is impacting human health*. The World Economic Forum. Retrieved November 29, 2023, from https://www.weforum.org/agenda/2023/11/health-day-climate-change-impacts-health-cop28/

Yeung, J. (2023, August 30). *Chinese people are living two years longer thanks to 'war on pollution,' report says*. CNN. Retrieved May 27, 2024, from https://edition.cnn.com/2023/08/30/asia/air-pollution-report-china-south-asia-intl-hnk-scn/index.html

CHAPTER 5

Building Materials and CO2 Emissions.

"When you discover you are riding a dead horse, the best strategy is to dismount" Dakota Native American Wisdom.

Life Cycle Assessment

Built environment-related emissions include embodied carbon from the materials and construction of the buildings and structures and operating carbon related to the use of the buildings from energy, waste and water. Life cycle assessment (LCA) is a tool for measuring the energy and other inputs throughout the full life cycle, or "cradle to grave" of the building (Bayer et al., 2010). LCA not only helps measure greenhouse gas (GHG) emissions at different stages of the entire building life from mineral extraction to operation and waste management stages of the life cycle. They also help identify potential reduction options (Reijinders, 2021). Social and economic impacts are excluded. Around 80% of GHG emissions are related to building operations and 20% to their construction (HSBC, 2025), although there are different figures offered from different sources. In particular the accounting for interior renovations and tenant fit outs over the lifetime of a building can vary considerably.

With the number of variables and complex calculations, standards are important to enable comparisons between manufacturers and to measure

S. Crispin, *Sustainability in the Built Environment*,
https://doi.org/10.1007/978-981-95-6549-8_5

progress in making changes. Even more so for certification, regulatory approvals, carbon tax accounting and so on. Requirements and guidelines for life cycle assessment (LCA) are covered by ISO14040. These include the scope and goals, inventory analysis, life cycle impact assessment, life cycle interpretation, reporting and review as well as limitations and relationship between phases and value choices (International Standards Organisation, 2006).

GHGs are the seven gases under the Kyoto Protocol that directly impact climate change: carbon dioxide (CO2), nitrous oxide (N2O), chlorofluorocarbons (CFCs), hydrofluorocarbons (HFCs), perfluorocarbons (PFCs), methane (CH4), sulphur hexafluoride (SF6) and nitrogen trifluoride (NF3) (UK NAEI, 2022). Different GHGs have different effects on the earth's warming depending on their ability to absorb energy and their lifetime. Global Warming Potential (GWP) measures the amount of energy the emissions of one tonne of a gas can absorb over its lifetime compared to one tonne of Carbon Dioxide (CO2) (US EPA, 2024).

The RICS published document "Whole Life Carbon Assessment for the Built Environment" sets out comprehensive professional guidance for undertaking a whole life carbon assessment of a building and includes considerations for building components (RICS, 2023).

The life cycle stages to be considered include:

1. Extraction and processing.
2. Transportation and manufacturing.
3. Transport to site.
4. Construction and installation.
5. Use & maintenance.
6. Repair, replacement and refurbishment.
7. Demolition.
8. Removal to waste disposal/recycling facility.
9. Recovery or waste processing.
10. Disposal (BSI, 2011).

According to IEA, GlobalABC and the United Nations Environment Programme, in 2022 the buildings and construction industry combined accounted for "37% of total energy-related emissions with almost 9% coming from materials used in the construction of buildings. In 2021

25% of building sector emissions came from embodied carbon and 75% from operational carbon." Under the business as usual scenario that ratio is expected to be close to 50:50 by 2050 (UNEP/GlobalABC, 2022). Changing the course of that increase in embodied carbon while reducing operational carbon are the keys to decarbonising the sector.

The next section considers the materials used in the commercial office property sector which accounts for few emissions than the residential sector but many of the inputs are the same.

Review of the Main Building Materials

We've all heard the term "concrete jungle", perhaps most relevant to New York City where of the over 12,000 miles of sidewalks, 90% are made of concrete, according to the Department of Transport. On their own, concrete buildings wouldn't last long. As a material concrete cracks under load or subject to the freeze–thaw cycle and it is affected by settling when the ground subsides (General Purpose Assistant Bot, 2019). To improve durability, concrete needs reinforcing, usually with steel but other additives improve the durability of the material too. Concrete itself is a composite material; according to the American Concrete Institute, it is a mixture of cement, aggregates, water and fibres, admixes and other materials (ACI, 2023). A steel and concrete structure is typically clad in a glass facade 'building envelope' to seal the interior. Partition walls, various plant and machinery are installed to manage a comprehensive building system, including air conditioning, vertical transportation (lifts and escalators) and then occupiers install interior decoration (fit out) materials, furniture, data rooms and other systems for fully functional use of the space they have bought or leased.

By material type, concrete and steel account for around 65% with brick at 18% and aluminium at 8% (Zhong et al., 2021). For this chapter, the materials in modern office buildings to be considered are cement, steel, concrete blocks, aluminium, wood, insulation and glass (Center for Sustainable Systems, University of Michigan, 2023). Understanding all the inputs to a building means knowing their carbon footprint. Given the differences in production of, say, steel in one country using coal power compared to steel in another country using green hydrogen energy sources, means knowing not just what the material is but where it has been produced. This is known as 'traceability' which is an essential part of developing sustainable supply chains that enable material inputs to be reduced and reused. This can lead to lower waste and costs (Saenz et al., 2021).

The following section takes a closer look at some of the different materials and a glimpse into their life cycles.

Cement

Cement is not a new material; buildings dating from ancient Rome that still stand today were built using it. What is new is the massive scale of production which has increased 30 fold since 1950 (Andrew, 2018). More than 4 billion tonnes of cement are made each year (Walker & Niranjan, 2022). The production of cement accounts for between 4 and 8% of global CO_2 emissions (Lehne & Preston, 2018) which results from the chemical reaction of the ingredients as limestone becomes lime and the fossil fuels used to generate the high temperatures of well over 1000 °C (Andrew, 2018). Limestone itself is a mineral made of calcium carbonate created over millions of years from the shells of marine creatures (Rosenwald, 2011).

Cement is a key input to make concrete alongside water and gravel which goes into everything from buildings, bridges, roads, dams. In buildings, concrete is used in pilings, raft foundations, columns, beams and slabs as well as being used to make blocks. That makes it one of the foundation materials of urbanisation but it sucks up further quantities of water and sand, worsens heat islands in cities and creates dust from both construction and demolition sites (Watts, 2019).

While the cement industry may be considered a conservative industry with limited scope for innovation, there has been an increase in cement-related patenting activity in recent years, mainly focused on clinker substitutes. In particular, low carbon technology patents have increased since 2008, with non-limestone-based binders showing the greatest potential (Lehne & Preston, 2018). As Andrew (2018) identified, emissions from cement production vary widely in different countries, so the mitigation efforts vary widely from clinker reduction or substitution, to magnesium-based cements. Despite all the patents, the right combination is yet to be found and scaled up.

The World Cement Association (WCA) represents cement producers worldwide, the corporate members. Associate members are made up of technology, equipment and service suppliers (WCA, n.d.). According to the WCA, cement is responsible for around 7% of the world's total CO_2

emissions from the thermochemical process of decomposing limestone at 1400 °C in a furnace or kiln. Burning fuel for clinker production accounts for around 30–35% of emissions and 65–70% from the decomposition of limestone (Riley, 2025). With no easy means to change that process and limited scope for alternatives at scale for traditional cement production, efforts can best focus on improving end-of-life treatment to increase circularity and perhaps reducing the increase in demand though design, construction methods and product selection.

There are several innovative uses for old concrete including spreading it on farmland. One Irish business is using small pieces of waste concrete to mimic the process by which rocks are weathered and capture carbon (carbon dioxide reacts with the concrete to form calcium carbonate, known as carbonation) and eventually sequester the carbon dioxide. A process that, at the same time, improves crop yields (Orf, 2024) by enhancing photosynthesis and improving nutrient uptake.

Steel

Like cement, steel production is a very carbon intensive process with blast furnace temperatures of 1400 °C. Each tonne of steel accounts for 1.5–2 tonnes of CO2, with the higher figure including indirect energy inputs, and around 7% of total CO2 emissions (IEA, 2021). Due to its affordability, availability, flexibility, strength and durability steel is widely used in buildings. In addition, it is readily recyclable (World Steel Association, n.d.). Overall, around 25% of all steel produced goes into building construction (Moynihan & Allwood, 2014). Energy in sufficient quantities for large-scale steel making can only be met from fossil fuels. This is beginning to change.

Not all steel has the same carbon footprint. Among other variables that account for differences in CO2 emissions is the source of energy used in different production locations. Energy for smelting is generally generated from coal, natural gas or electricity. In 2019, steel production used 8% of total energy supplies and contributes 10–25% of the total product cost (IEA, 2021). The material extraction stage of the life cycle is less energy intensive. Iron ore extraction and processing generates just under 12 kg of CO2 for 1 tonne of iron ore, of which 50% comes from loading and hauling (Haque & Norgate, 2015), that is the massive trucks running on diesel used to transport the ore.

The decarbonisation of steel production is still at the early stages. According to the World Economic Forum, green steel is being made from hydrogen instead of fossil fuel energy by Swedish manufacturer SSAB. One leading example is the Hydrogen Breakthrough Ironmaking Technology (Hybrit), a joint venture between SSAB, a mining company LKAB and a power company Vattenfall. Hybrit has even set a net zero target for 2030, reducing CO2 emissions by 8 million tonnes (SSAB, 2022). This of course depends on green hydrogen, currently very expensive, rather than that produced using fossil fuels.

Other than using hydrogen, the other means of producing low carbon steel include carbon capture, a high-investment approach, or using electricity (Rathi, 2022). Boston Metal uses Molten Oxide Electrolysis (MOE) to decarbonise steelmaking. Rather than try to explain it in my own words, here is a direct quote from the Boston Metal website:

"We've developed an electrolysis process that eliminates the need for coal in steel production. Our MOE platform uses renewable electricity to convert all iron ore grades to high-quality liquid metal. This direct approach eliminates several steps in the steelmaking process and does not require coke production, iron ore sintering and pelletizing, blast furnace reduction or basic oxygen furnace refinement.

In the MOE cell, an inert anode is immersed in an electrolyte containing iron ore, and then it's electrified. When the cell heats to 1600 °C, the electrons split the bonds in the iron oxide in the ore, producing pure liquid metal. No carbon dioxide or other harmful by-products are generated, just oxygen. Furthermore, MOE does not require process water, hazardous chemicals or precious-metal catalysts" (Boston Metal, n.d.).

While decarbonising the production process is good news, it is insufficient. Increasing circularity and developing scrap-based production from 20% in 2020 to 27% by 2030 and 40% by 2050 is essential to effectively decarbonise the industry (Vass et al., 2021).

Concrete Blocks

Concrete blocks, also known as breeze or cinder blocks, are used in almost every building. They are made of cement, water, coarse aggregates, fine aggregates, sodium silicate and sodium hydroxide (Damen et al., 2017, 1625). Additional inputs are energy for transport, compaction and curing. Studies in China show that each tonne of concrete block produces around

97 kg of CO2, of which 90 kg is from the cement (Liu et al., 2018, 4). Concrete blocks, including paving blocks, come in various types which we do not need to detail here, but are generally hollow which reduces weight, improves thermal and noise insulation and reduces material inputs. They provide a quick way to speed up the construction process because they are bigger than traditional clay bricks (Singh, n.d.).

Efforts to reduce the carbon footprint of concrete blocks can have surprising results. One means to achieve a more sustainable block might be to use recycled aggregates (RA) instead of natural aggregates (NA). The Liu et al. study showed that RA produces 57% less CO2 than NA, but using them in concrete blocks can result in increased CO2 emissions due to the need to use 64% more cement to deliver a block of the same strength. The CO2 benefit of saving RA from landfill is outweighed by the additional cement needed and the emissions from NA production is relatively small. Elsewhere, 10 years of research at Heriot-Watt University and different techniques have resulted in unfired bricks made from 90% construction waste, resulting in a product that has less than a tenth of the carbon emissions of a traditional brick (Diamond, 2022). Other innovative materials being used to produce low carbon bricks include loofah, charcoal, urine, carrot nanoparticles and mushroom mycelium (Aouf, 2019).

Jaafar et al (2023) looked at using a range of waste materials as alternatives to coarse aggregates to produce sustainable concrete blocks. Materials considered include crushed clay bricks, crushed concrete, plastic waste materials and crushed thermos stone. The study found that different aggregate replacements can alter the properties of the block in different ways, making them suitable for different purposes. Demolition waste materials are the readiest replacement for natural aggregates while plastic waste materials are more suitable for paving blocks because they can increase compressive strength. Natural aggregates help improve porosity of blocks in situations where that is a required property. Whichever replacement material is used, care needs to be taken to find the optimum proportion of RA to NA while maintaining the required technical characteristics of the product (Jaafar et al., 2023).

Timber

Timber, in various forms, has historically been used in buildings and many of the most admired historical buildings are wooden structures. It is a renewable material which absorbs CO2 from the atmosphere and, usefully, stores carbon during its service life, only giving off CO2 if disposed of in landfill or by incineration (Ramasamy et al., 2020).

Taking raw timber as the baseline product, energy is used in production, transportation and on disposal. CO2 released at end of life is equal to that absorbed during the growth stage; this is known as a carbon neutral product. Manufactured wood products include laminated timber, plywood, particleboard and fibreboard. Notably, plywood is a composite material that has been used in the manufacture of planes (aircraft grade), useful for evading radar detection. Marine grade plywood is used in, among other things, sailing dinghies, useful for messing around in and easy to repair.

One of the more exciting timber products for construction is cross-laminated timber (CLT) which has become popular in Europe (Souza, 2018) and being seen now in high rise building projects in the United States (Gu et al., 2021, 1). Being a carbon neutral product, wood would seem to be an easier nut to crack than other building materials. Of course it isn't that simple. The answer is sustainable forestry, which the Rainforest Alliance define as "the extent to which forestry practices mimics natural patterns of disturbance and regeneration" (Rainforest Alliance, 2016). This means balancing the needs of local communities, wildlife and the environment and involves certification by the Forest Stewardship Council (FSC), founded by the Rainforest Alliance.

CLT has been used in a range of building functions and has been shown to offer structural load-bearing capacity and fire resistance as well as being able to deal with environments like swimming pools. T3 Collingwood is a development in Melbourne being built by Hines, a global real estate, investment, development and property manager. T3 stands for timber, transit and technology and the 15-storey, 18,200 sq.m building said to reduce embodied carbon by around 34% under the Australian Green Star methodology (Hines, 2023).

Aluminium

The production of aluminium starts with mining bauxite; it is then processed into alumina before entering an electrolytic smelting process and going on to casting. The industry accounts for around 2% of total greenhouse gas emissions. More than 95% of electricity in the ore to ingot stage is used for smelting (Nunez & Jones, 2016). Ingots go on for further processing into various materials to be used in buildings.

The significance of the energy input into smelting means that the source of energy is relevant in determining the carbon footprint of the ingot. China produces around 60% of new aluminium and Chinese aluminium smelting uses 90% coal-produced energy while in Europe over 76% comes from hydropower (Nunez & Jones, 2016, 6). Production in Gulf Cooperation Countries in contrast, mostly use natural gas (WEF, 2020). There are signs that national net zero targets and consumer demand are having an influence. In 2019, China Hongqiao Group, a major private sector aluminium producer perhaps mindful of China's 2030 peak carbon emissions targets, announced plans to relocate 2 million tonnes of smelting capacity from coal rich Shandong province to hydropower-rich Yunnan province. This was followed in 2021 by an announcement for a further 1 million tonnes capacity to move, a combined nearly 50% of the company's total production (Daly, 2021).

Happily, aluminium is virtually infinitely recyclable and is already one of the most recycled materials (The Aluminium Association, n.d.). The Aluminium Association estimates that around 75% of all the aluminium ever produced is still being used and recycling the metal only uses 5% of the energy needed to make new aluminium (The Aluminium Association, 2017). As with steel, recycling is a critical part of reducing emissions.

Another critical change, according to Mission Possible Partnership, is for stakeholders in the value chain to collaborate. Higher prices are inevitable, but with users also seeking to reduce their carbon footprints, this is a possibility as demand for green aluminium increases. This in turn encourages banks to fund green production through green loans, green bonds and other instruments. Government grants or tax incentives for smelter-retrofitting and perhaps carbon-pricing mechanisms would also help the transition (Assunção et al., 2023).

Plasterboard

Plasterboard, also known as drywall, is made of gypsum, paper and various fibrous materials and is used as an alternative to wet plaster (Johal, 2021). The main environmental impact from gypsum plasterboard is the energy in manufacture and transportation (Zhang et al., 2020).

The Gypsum Products Development Association (GPDA) represents 4 major UK and Ireland-based gypsum board and plaster manufacturers. It claims that gypsum is eternally recyclable and that this is enabling the "construction chain towards a positive environmental footprint". Being eternally recyclable is not the same as actually being recycled. According to GPDA, new plasterboard contains 11.5% post-consumer recycled content (GPDA, 2024). Post-consumer recycled content is material that is repurposed after consumer use, not the same as pre-consumer recycled content which would be scraps from the production or processing stage (Chuyko, 2023). Progress is important and the industry has taken steps to improve sustainability performance at extraction and processing stages. It has set carbon reduction targets in manufacturing, increasing recycled content rates, reducing on site waste, improving design for deconstruction, using natural gypsum from flue gas desulphurisation of the power station industry (FGD Gypsum), which has a higher purity level than natural gypsum, according to the GDPA.

Looking at the academic research, (Morsali et al., 2024) use a life cycle assessment in accordance with ISO14040 to measure a range of impact categories in UK plasterboard production:

1. Global warming potential (GWP).
2. Stratospheric ozone depletion.
3. Ionising radiation.
4. Ozone formation (human health).
5. Fine particulate matter formation.
6. Ozone formation (terrestrial ecosystems).
7. Terrestrial acidification.
8. Freshwater eutrophication.
9. Marine eutrophication.
10. Terrestrial ecotoxicity.
11. Human carcinogenic toxicity.
12. Human non-carcinogenic toxicity.
13. Land use.

14. Mineral resource scarcity.
15. Fossil resource scarcity.
16. Water consumption.

This list helps to show the complexity of measuring impact, in particular when the research confirms that the greatest impact is human carcinogenic toxicity. This comes not only from the gypsum production itself but also from the manufacture of steel screws used in plasterboard installation. Fastening is within the LCA boundary; happily alternative fastening solutions provide a relatively easy fix.

Lower impact plasterboard solutions includes a transition to renewable energy in manufacturing, improved recycling and waste management, optimising logistics to reduce emissions.

Insulation Materials

Insulation materials are installed to reduce energy consumption during the operational phase of a building life cycle and are typically made of mineral wool or polyurethane and extruded polystyrene (Llantoy et al., 2020). Mineral Wool is made from stone or silica heated to a molten state and then spun into fibres that form a thick carpet or mat (Mineral Wool Manufacturers Association, n.d.). Polyurethane comes in panel and spray forms. The raw materials (including a petroleum by-product and polyol), manufacture, transportation and disposal stages contribute the vast majority, around 95%, of CO2 equivalent emissions during the life cycle of a sample product (PU Europe, n.d.).

Post-consumer clothing and paper can also be used to make insulation materials. Using bio-based insulation materials can significantly reduce emissions compared to the materials described above. Bio-materials include cotton, wool, straw, flax, hemp, wood and other natural fibres. While they need treatment for fire resistance, that is no different to man-made materials which can be notoriously difficult to keep fire-safe (Raja et al., 2023).

Biochar is an interesting material to incorporate into insulation materials. It is a plant waste material similar to charcoal that retains the carbon absorbed from the plant rather than releasing it when heated in a kiln or similar and deprived of oxygen (Ghoniem et al., 2024). Making use of agricultural waste to produce biochar to add to insulation and other materials locks away the CO2.

Another plant material for insulation is hemp. Hemp sequesters up to 2 tonnes of CO_2 per tonne of fibre (Fan & Fu, 2017). US firm Hempitecture makes bio-based insulation products including Hempwool, Fibrepad and Hempcrete (Hemptecture, 2024).

Like hemp, bamboo is another fast-growing plant. It can be harvested within 5 years compared to 20–40 years for timber and has similar thermal properties to wood (Wang et al., 2017). It is a familiar sight on Asian construction sites used as a support or scaffolding, it is also a natural insulator (Nguyen, 2017).

Glass

There are different types of glass and different manufacturing processes for glass bottles and glass windows. Windows are made from a type of flat glass called float glass which is made by "floating" the molten material on molten tin, before being moved on to ceramic rollers to cool (Hill, 2019). This manufacturing process results in a smooth and distortion-free product that can then be laminated, heat-toughened, coated or otherwise treated (Asahi India Glass, 2019). The raw materials that go into float glass include limestone, sand, dolomite, soda ash and glass cullet.

Sand is the main material and, in Europe, gas is the main energy source with oil for transportation; water is another significant input. CO_2 is the main emission and 1 kg of glass produces around 1.23 kg of CO_2 (Glass for Europe, 2011).

Melting glass cullet doesn't release CO_2 in the same way melting the raw materials does, but flat glass used in windows cannot contain cullet impurities unless it too is derived from flat glass (Nature, 2021).

Glass for Europe is a trade association for Europe's flat glass producers and their suppliers. Their report "Flat Glass in a Climate-Neutral Europe - Triggering a Virtuous Cycle of Decarbonisation" states that as much as 37% of total energy consumption in EU buildings could be saved in 2050 with high-performing glazing products. Someone has some new facades to sell. The case is compelling, combining a range of technologies such as building integrated photovoltaics (BIPV), transparent photovoltaics, switchable glazing and making windows that are more durable all help offset the energy intensity of the manufacturing process (Glass for Europe, 2020). Reducing the impact of glass manufacturing is a matter of switching to renewable energy sources; as with cement and steel, this is not yet commercially viable. The other problem is that most flat glass exports (13% in 2021) are made in China (Jaganmohan, 2024) using more carbon-intense methods, 90% more CO2 per tonne of flat glass according to Glass for Europe based on figures from the China Architectural & Industrial Glass Association 2014 Annual Conference.

Global Warming Potential

For each of these materials, it is possible to calculate total GHG emissions by breaking down each individual material associated with the extraction and manufacturing process and its global warming potential (GWP) relative to CO2, which is rated at 1. Once the GWP multiplier is applied to each material a CO2 equivalent is obtained. The cumulative total of all the CO2 equivalents is the carbon footprint or total embodied carbon released in the manufacture of each product (Damen et al., 2017, 1630). This in turn provides the building blocks for assessing the total carbon footprint of an entire building's embodied carbon.

In addition, energy consumed on site during the construction process should be included. This includes temporary accommodation provided for workers, site offices, water and wastewater as well the impact of all other waste generated and its disposal.

Operating carbon is generated by the building components using energy, being maintained, repaired, refurbished or replaced (RICS, 2017). A particular source of GHG emissions is the air conditioning system including energy, refrigerants and refrigerant leaks, HVAC system insulating materials and ducting. Other sources include cleaning materials, staff uniforms and water. This will be covered in some detail in Chapter 7.

End-of-life stage of a building is reached when it is no longer used and it is dismantled and demolished and is considered complete when the site is clear and level (RICS, 2017, 30). At end of life, materials may be reused on site, reused elsewhere, sent for recycling off-site or disposed of in landfill or by incineration. Emissions related to these pathways are included in the end-of-life assessment.

What Are the Greatest Environmental Impacts from Buildings?

As has been discussed, the carbon footprint of a building comprises embodied and operational carbon. Studies show that building information modelling (BIM) can be used to effectively analyse embodied carbon and predict operating carbon based on life cycle energy simulation. Such modelling can be used to optimise buildings based on structural design, low carbon materials, and innovative construction methods. (Gan et al., 2023, 2).

Of the materials covered above, 23% of total global emissions come from only 3 of them, steel, concrete and aluminium, all of which are used in the built environment (Architecture 2030, n.d.).

What is often overlooked is the impact of the interiors. In parts of Asia, lease terms for an office unit can be as short as 3 years, with the office unit handed over to the tenant as a bare shell every time the tenant changes. The previous tenant is usually responsible for restitution on their lease expiry. This means that, potentially, a whole new interior renovation or fit out can be required every 3 years. The impact is even greater in shopping centres, where fit out carbon from retail interiors is approximately twice that of office interiors, according to RESET whose embodied carbon standard is used to measure and manage embodied carbon for interiors and the base building (RESET, 2024). There is further discussion on interior fit outs in Chapter 7.

Summary

This chapter has provided insight into the complexity of the subject matter. What remains important, while acknowledging that complexity, is that property professionals can safely leave that to academics and scientists while remembering that the built environment is critical to both human wealth and health. We must all do everything we can to optimise sustainability in materials, construction, use and end-of-life demolition stages of the life cycle.

Decarbonising the material inputs that go into buildings is a work in progress and that is where Mission Possible Partnership is driving the transition "across the entire value chain of the world's highest-emitting heavy industry and transport sectors" (Mission Possible Partnership, 2024). Much depends on their success.

References

ACI. (2023, January 6). *Definition of concrete, hydraulic cement, mortar, and grout*. American Concrete Institute. Retrieved May 29, 2024, from https://www.concrete.org/tools/frequentlyaskedquestions.aspx?faqid=640

Andrew, R. M. (2018). Global CO2 emissions from cement production. *Earth System Science Data, 10*, 195–217. https://doi.org/10.5194/essd-10-195-2018

Aouf, R. S. (2019, July 14). *Bricks made from loofah and charcoal could promote biodiversity in cities*. Dezeen. Retrieved June 21, 2022, from https://www.dezeen.com/2019/07/14/green-charcoal-bricks-loofah-technology-materials/

Architecture 2030. (n.d.). *Why the building sector? – Architecture 2030*. Architecture 2030. Retrieved May 22, 2023, from https://architecture2030.org/why-the-building-sector/

Asahi India Glass. (2019, December 22). *Everything you need to know about float glass*. AIS Glass. Retrieved May 15, 2022, from https://www.aisglass.com/all-you-need-to-know-about-float-glass/

Assunção, P., Lorch, J., Olson, K., Penso, R., Stürtz, M., & UlanoV, A. (2023, April 20). *Sustainable aluminum: Decarbonizing at a cost that makes sense*. McKinsey. Retrieved May 31, 2024, from https://www.mckinsey.com/industries/metals-and-mining/our-insights/aluminum-decarbonization-at-a-cost-that-makes-sense/

Bayer, C., Gamble, M., Gentry, R., & Joshi, S. (2010). *A guide to life cycle assessment of buildings* [Assessment in practice]. The American Institute of Architects. Retrieved 5 10, 2022, from https://www.aia.org/resources/7961-building-life-cycle-assessment-in-practice

Boston Metal. (n.d.). *Green steel solution*. Boston Metal. Retrieved May 31, 2024, from https://www.bostonmetal.com/green-steel-solution/

BSI. (2011). *BS EN15978:2011* [Sustainability of construction works. Assessment of environmental performance of buildings. Calculation method]. European Standards. Retrieved 2022, from https://www.en-standard.eu/bs-en-15978-2011-sustainability-of-construction-works-assessment-of-environmental-performance-of-buildings-calculation-method/

Center for Sustainable Systems, University of Michigan. (2023). *Commercial Buildings Factsheet | Center for Sustainable Systems*. Center for Sustainable Systems. Retrieved May 29, 2024, from https://css.umich.edu/publications/factsheets/built-environment/commercial-buildings-factsheet

Chuyko, S. (2023, August 23). *Certified circular: What is post-consumer recycled content certification?* Sustainable Jungle. Retrieved May 31, 2024, from https://www.sustainablejungle.com/certifications/recycled-content-certification

Daly, T. (2021, May 14). *Aluminium giant China Hongqiao to move more smelting capacity to Yunnan - sources*. Reuters. Retrieved May 31, 2024, from https://www.reuters.com/article/idUSL4N2N101W/

Damen, J., Kim, J. C., & Ouellet-Plamondon, C. (2017). Life cycle assessment of emergent masonry blocks. *Journal of Cleaner Production, 171*. https://doi.org/10.1016/j.jclepro.2017.10.044

Diamond, A. (2022, February 8). *Decarbonising cement by 2040: University, government and industry collaboration unveiled*. Heriot-Watt University. Retrieved June 21, 2022, from https://www.hw.ac.uk/news/articles/2022/decarbonising-cement-by-2040-university.htm

Fan, M., & Fu, F. (2017, November 9). *A perspective – Natural fibre composites in construction*. Retrieved May 31, 2024, from https://www.sciencedirect.com/topics/engineering/natural-insulation-material

Gan, J., Li, K., Li, X., Mok, E., Ho, P., Law, J., Kwok, R., & Yau, R. (2023). Parametric BIM-Based Lifecycle Performance Prediction and Optimisation for Residential Buildings Using Alternative Materials and Designs. *Buildings, 13*(4), 904. https://doi.org/10.3390/buildings13040904

General Purpose Assistant Bot. (2019, August 16). *How long will un-reinforced concrete last? I want to cover certain parts of my yard, the ground, with concrete as a way to limit the growth of grass, no steel nor rebar of any kind*. Quora. Retrieved May 29, 2024, from https://www.quora.com/How-long-will-un-reinforced-concrete-last-I-want-to-cover-certain-parts-of-my-yard-the-ground-with-concrete-as-a-way-to-limit-the-growth-of-grass-no-steel-nor-rebar-of-any-kind

Ghoniem, A., Crane, R. C., & Krol, A. (2024, February 27). *Biochar*. MIT climate portal. Retrieved May 31, 2024, from https://climate.mit.edu/explainers/biochar

Glass for Europe. (2011). *Report - Life cycle assessment - Float glass LCA*. Glass for Europe. Retrieved May 15, 2022, from https://glassforeurope.com/report-life-cycle-assessment-of-float-glass/

Glass for Europe. (2020). *Flat glass in climate-neutral Europe*. Glass for Europe. Retrieved May 31, 2024, from https://glassforeurope.com/wp-content/uploads/2020/01/flat-glass-climate-neutral-europe.pdf

GPDA. (2024). *Sustainability*. Gypsum Products Development Association. Retrieved May 31, 2024, from https://gpda.com/sustainability/

Gu, H., Liang, S., Pierbon, F., Puettmann, M., Ganguly, I., Chen, C., Pasternack, R., Wishie, M., Jones, S., & Maples, I. (2021). Mass timber building life cycle assessment methodology for the U.S. regional case studies. *Sustainbility*, *13*. https://doi.org/10.3390/su132414034

Haque, N., & Norgate, T. (2015). *Iron ore: Mineralogy, processing and environmental sustainability*. Elsevier Science. https://publications.csiro.au/rpr/pub?pid=csiro:EP1311257

Hemptecture. (2024). *About Hempitecture*. Hempitecture: Home. Retrieved May 31, 2024, from https://www.hempitecture.com/

Hill, L. (2019, August 2). *Float glass vs. Flat glass vs. Plate glass vs. Sheet glass*. Glass.com. Retrieved May 15, 2022, from https://info.glass.com/float-flat-plate-sheet-glass/

Hines. (2023, May 26). *New Collingwood timber office building to reduce carbon emissions by….* Hines. Retrieved May 31, 2024, from https://www.hines.com/news/new-collingwood-timber-office-building-to-reduce-carbon-emissions-by-34-percent-compared-to-standard-concrete-and-steel-construction

HSBC. (2025). *Decarbonising the built environment* [Perspectives from mainland China and Hong Kong].

International Standards Organisation. (2006). *ISO 14040:2006 - Environmental management — Life cycle assessment — Principles and framework.* ISO. Retrieved May 29, 2024, from https://www.iso.org/standard/37456.html

International Energy Agency. (2021, February). *Energy technology perspectives 2020.* Retrieved March 1, 2024, from https://iea.blob.core.windows.net/assets/7f8aed40-89af-4348-be19-c8a67df0b9ea/Energy_Technology_Perspectives_2020_PDF.pdf

Jaafar, A., Zena, A., & Allawi, A. (2023, September 10). *Studying sustainable concrete block efficiency production: A review.* ResearchGate. Retrieved May 31, 2024, from https://www.researchgate.net/publication/373636804_Studying_Sustainable_Concrete_Block_Efficiency_Production_A_Review

Jaganmohan, M. (2024, May 22). *Glass industry worldwide - Statistics & facts.* Statista. Retrieved May 31, 2024, from https://www.statista.com/topics/4108/glass/#topicOverview

Johal, J. (2021, October 11). *What is plasterboard and how is it made? A simple guide.* Building Materials. Retrieved May 4, 2022, from https://www.buildingmaterials.co.uk/nuts-and-bolts/guides/what-is-plasterboard-and-what-is-it-used-for

Lehne, J., & Preston, F. (2018). *Making concrete change* [Innovation in low-carbon cement and concrete]. Chatham House. Retrieved May 10, 2022, from https://www.chathamhouse.org/2018/06/making-concrete-change-innovation-low-carbon-cement-and-concrete

Liu, Y., Hossain, M. U., & Ling, B. T.-C. (2018). Carbon footprint of block prepared with recycled aggregate: a case study in China. *IOP Conference Series: Materials Science and Engineering, 431.* https://doi.org/10.1088/1757-899X/431/3/032009

Llantoy, N., Chàfer, M., & Cabeza, L. F. (2020). A comparative life cycle assessment (LCA) of different insulation materials for buildings in the continental Mediterranean climate,. *Energy and Buildings, 225.* https://doi.org/10.1016/j.enbuild.2020.110323

Mineral Wool Manufacturers Association. (n.d.). *What is mineral wool?* MIMA. Retrieved May 4, 2022, from http://mima.info/about-mineral-wool/what-is-mineral-wool/

Mission Possible Partnership. (2024). *Mission possible partnership.* Home. Retrieved May 31, 2024, from https://www.missionpossiblepartnership.org/

Morsali, S., Rakhshanbabanari, K., Osmani, M., Cavalaro, S., Gutai, M., Castro-Diaz, M., Parker, B., Sparkes, J., Needham, P., & Newport, S. (2024). Life cycle assessment of plasterboard production: A UK case study. *Sustainability, 16*. https://doi.org/10.3390/su16093564

Moynihan, M. C., & Allwood, J. M. (2014). Utilization of structural steel in buildings. *Proceedings of the Royal Society, 470*. https://doi.org/10.1098/rspa.2014.0170

Nature. (2021, November 3). *Glass is the hidden gem in a carbon-neutral future*. Nature. Retrieved May 31, 2024, from https://www.nature.com/articles/d41586-021-02992-8

Nguyen, D. (2017, November 9). *Know how bamboo can be used as an insulator?* Bambooye. Retrieved May 31, 2024, from https://bambooye.com/bamboo-as-an-insulator/

Nunez, P., & Jones, S. (2016). Cradle to gate: Life cycle impact of primary aluminium production. *The International Journal of Life Cycle Assessment, 21*, 1594–1604. https://doi.org/10.1007/s11367-015-1003-7

Orf, D. (2024, April 25). *A new use for old concrete could revolutionize carbon capture*. Popular Mechanics. Retrieved May 30, 2024, from https://www.popularmechanics.com/science/environment/a60537619/a-new-use-for-old-concrete-could-revolutionize-carbon-capture/

PU Europe. (n.d.). *What is polyurethane? – Excellence in Insulation*. Excellence in Insulation. Retrieved May 4, 2022, from https://highperformanceinsulation.eu/home/insulating-with-pu/what-is-polyurethane/

Rainforest Alliance. (2016, July 28). *What is sustainable forestry?* Rainforest alliance. Retrieved June 21, 2022, from https://www.rainforest-alliance.org/insights/what-is-sustainable-forestry/

Raja, P., Muruagan, V., Ravichandran, S., Behera, L., Mensah, R. A., Mani, S., Kasi, A., Balasubramanian, K. B. N., Sas, G., Vahabi, H., & Das, O. (2023, January 6). *A review of sustainable bio-based insulation materials for energy-efficient buildings*. Wiley. Retrieved May 31, 2024, from hthttps://onlinelibrary.wiley.com/doi/full/10.1002/mame.202300086

Ramasamy, G., Gan, K., & Zairul, A. (2020). *Concept of life cycle assessment (LCA) in timber industry*. Forest Research Institute Malaysia. Retrieved April 16, 2022, from https://www.researchgate.net/publication/351769960_CONCEPT_OF_LIFE_CYCLE_ASSESSMENT_LCA_IN_TIMBER_INDUSTRY

Rathi, A. (2022, October 6). *Inside the Bill Gates-backed startup making zero-emissions steel*. Bloomberg. Retrieved May 31, 2024, from https://www.bloomberg.com/news/features/2022-10-06/how-to-make-emissions-free-iron-at-temperatures-colder-than-coffee?sref=jjXJRDFv

Reijinders, L. (2021). Life Cycle Assessment of Greenhouse Gas Emissions. *Handbook of Climate Change Mitigation and Adaptation*. https://doi.org/10.1007/978-1-4614-6431-0_2-3

RESET. (2024, November 22). *RESET Factoid #3 - Embodied carbon for retail interiors vs. Core & Shell | RESET®*. RESET® Standard. Retrieved June 21, 2025, from https://www.reset.build/newsroom/453

RICS. (2017). Whole life carbon assessment for the built environment. *RICS Professional Standards and Guidance, UK*. rics.org/guidance

RICS. (2023). *Whole life carbon assessment for the built environment*. RICS. Retrieved June 21, 2025, from https://www.rics.org/content/dam/ricsglobal/documents/standards/Whole_life_carbon_assessment_PS_Sept23.pdf

Riley, I. (2025, February). Cement demand forecast 2050. *Global Cement Magazine*, 12–18.

Rosenwald, M. (2011). *Building a better world with green cement | Science*. Smithsonian Magazine. Retrieved May 10, 2022, from https://www.smithsonianmag.com/science-nature/building-a-better-world-with-green-cement-81138/

Saenz, H., Hinkel, J., & Bysong, T. (2021, September 30). *Traceability: The next supply chain revolution*. Bain & Company. Retrieved April 19, 2022, from https://www.bain.com/insights/traceability-the-next-supply-chain-revolution/

Singh, K. (n.d.). *Concrete blocks - Types, uses, advantages & disadvantages*. Civil Engineering Portal. Retrieved May 31, 2024, from https://www.engineeringcivil.com/concrete-blocks-types-uses-advantages-disadvantages.html

Souza, E. (2018, May 20). *Cross laminated timber (CLT): What it is and how to use it*. ArchDaily. Retrieved April 16, 2022, from https://www.archdaily.com/893442/cross-laminated-timber-clt-what-it-is-and-how-to-use-it

SSAB. (2022). *Timeline for HYBRIT and fossil-free steel*. SSAB. Retrieved May 11, 2022, from https://www.ssab.com/en/company/sustainability/sustainable-operations/hybrit-phases

The Aluminium Association. (2017, November 9). *Infinitely recyclable*. Retrieved May 31, 2024, from https://www.aluminum.org/Recycling

The Aluminium Association. (n.d.). *Sustainability – Recycling | Aluminum Association*. The Aluminum Association. Retrieved April 19, 2022, from https://www.aluminum.org/Recycling

UK NAEI. (2022, October 14). *Overview of greenhouse gases - NAEI, UK*. National Atmospheric Emissions Inventory. Retrieved May 29, 2024, from https://naei.beis.gov.uk/overview/ghg-overview

UNEP/GlobalABC. (2022). *Sustainable building materials hub*. GlobalABC. Retrieved June 21, 2025, from https://globalabc.org/index.php/sustainable-materials-hub/home

US EPA. (2024, March 27). *Understanding global warming potentials | US EPA*. Environmental Protection Agency. Retrieved May 29, 2024, from https://www.epa.gov/ghgemissions/understanding-global-warming-potentials

Vass, T., Mandová, H., Gouy, A., & Levi, P. (2021, November). *Iron and steel – Analysis*. IEA. Retrieved June 21, 2022, from https://www.iea.org/reports/iron-and-steel

Walker, T., & Niranjan, A. (2022, February 4). *Green construction: Fixing concrete's carbon footprint | Environment | All topics from climate change to conservation*. DW. Retrieved April 6, 2022, from https://www.dw.com/en/concrete-cement-climate-carbon-footprint/a-60588204

Wang, J., Demartino, C., Xiao, Y., & Li, Y. Y. (2017, November 9). *Thermal insulation performance of bamboo- and wood-based shear walls in light-frame buildings*. Science Direct. Retrieved May 31, 2024, from https://www.sciencedirect.com/science/article/abs/pii/S0378778817305960

Watts, J. (2019, February 25). Concrete: the most destructive material on Earth | Cities. *The Guardian*. https://www.theguardian.com/cities/2019/feb/25/concrete-the-most-destructive-material-on-earth

WCA. (n.d.). *WCA who are we?* World Cement Association: WCA. Retrieved May 30, 2024, from https://www.worldcementassociation.org/

WEF. (2020). *Aluminium for climate: Exploring pathways to decarbonize the aluminium industry*. www3.weforum.org. Retrieved May 31, 2024, from https://3stepsolutions.s3-accelerate.amazonaws.com/assets/custom/010856/downloads/WEF_Aluminium_for_Climate_2020.pdf

World Steel Association. (n.d.). *Buildings and infrastructure - worldsteel.org*. World Steel Association. Retrieved April 10, 2022, from https://worldsteel.org/steel-by-topic/steel-markets/buildings-and-infrastructure/

Zhang, Y., Ma, L., Ren, S., Huang, M., Wang, Y., & Zhang, Q. (2020). Comparative life cycle assessment between ordinary gypsum plasterboard and functional phase-change gypsum plasterboard. *Materials Science Forum, 993*, 1473–1480. https://doi.org/10.4028/www.scientific.net/msf.993.1473

Zhong, X., Hu, M., Deetman, S., Steubing, B., Lin, H., Hernandez, G. A., Harpprecht, C., Zhang, C., Tukker, A., & Behrens, P. (2021, October 21). Global greenhouse gas emissions from residential and commercial building materials and mitigation strategies to 2060. *Nature Communications*. https://doi.org/10.1038/s41467-021-26212-z

CHAPTER 6

Decarbonising Building Materials

"I cannae change the laws of physics!" Montgomery "Scotty" Scott (Stardate 1704.2).

Abatement

Abatement is the term used to describe an intervention to reduce GHG emissions. Production of materials with high GHG emissions, such as energy intensive production processes, are known as hard to abate materials. Abatement cost is the cost required to reduce GHG emissions by 1 tonne, for example removing all fossil fuel burning equipment from a building and replacing with fully electric alternatives (Hallegatte, 2023).

Indirect emissions outside an organisations direct control are known as Scope 3, as defined by the GHG Protocol. These cover both upstream and downstream emissions related to the value chain and include:

1. Purchased goods and services.
2. Business travel.
3. Employee commuting.
4. Waste disposal.
5. Use of sold products.
6. Transportation and distribution.

S. Crispin, *Sustainability in the Built Environment*,
https://doi.org/10.1007/978-981-95-6549-8_6

7. Investments.
8. Leased assets and franchises (Carbon Trust, n.d.).

To recap, the building life cycle is summarised as:

1. Extraction of raw materials.
2. Processing and manufacturing.
3. Construction.
4. Use and refurbishment.
5. End of life.

Reducing Embodied Carbon

According to a report by the World Business Council for Sustainable Development and Arup, a development consultancy, embodied carbon can be reduced by 50% by taking the following steps:

1. Data led decision-making.
2. Setting carbon budgets.
3. Reducing knowledge gaps to optimise carbon reductions.
4. Collaborating across the supply chain.
5. Acting with urgency (WBCSD, 2023).

Another view, with a focus on Hong Kong and mainland China, published by HSBC, endorses much of this. Property developers are unable to set a carbon budget because they do not have a clear understanding of material production and measurement of GHG emissions and cannot come to agreement on carbon accountability; who is responsible for what emissions? Resolving this requires government leadership in including carbon requirements in public tendering for infrastructure projects, clearer carbon accountability and data standards and cross-border knowledge sharing between mainland China and Hong Kong to encourage low carbon building materials supply chain (HSBC, 2025).

The most useful insights come from looking at what some property developers and investors are doing. Hines is one of the largest privately held real estate investors and managers in the world (Hines, n.d.). In April

2022 they published the Embodied Carbon Reduction Guide which identifies structural materials as the largest contributor to embodied carbon and sets out building components as follows:

1. Structure
   a. Concrete
   b. Steel
   c. Timber
2. Envelope
   a. Cladding
   b. Insulation
3. Finishes
   a. Drywall
   b. Metal studs
   c. Carpet
   d. Ceiling Tiles
4. Other
   a. Elevators
   b. Mechanical, Electrical and Plumbing (MEP) equipment
   c. Building maintenance units (BMUs) (Hines, 2022).

Of the materials listed above, on a "cradle to site" basis, analysis of One Taikoo Place, an office building in Hong Kong, meaning extraction to site delivery, shows that carbon emissions from concrete account for 32.6%, rebar 52.8% and structural steel 3.8% (Gan et al., 2023). These proportions will vary from project to project, depending on the inputs, notably the carbon intensity of the energy used in the production stage and the distance and means of transport.

Absolute Zero by 2040 is a document by Lendlease for their Scope 3 Emissions Protocol. Lendlease is an integrated real estate business covering investments, development and construction (Lendlease, n.d.). According to their published materials, Lendlease Scope 3 emissions account for 90% of their global carbon footprint and the company target includes "eliminating Scope 3 emissions within the Lendlease defined boundary without the use of offsets" by 2040. The defined boundary for building materials they use includes cradle-to-gate emissions (extraction, production and transportation), while for use of a sold product such as an apartment, that would include the energy consumption during the in-use phase of the life of the apartment.

Hang Lung Properties is a property developer with headquarters in Hong Kong and projects in Hong Kong and mainland China. In March 2025 they published "Our Journey to Net Zero. Our Scenarios and Actions to Reduce Greenhouse Gas Emissions to 2050." Launching the paper the company stated: "Our paper highlights overlooked emissions, discusses carbon accounting challenges, urges greater transparency, and advocates for standardized benchmarks in construction materials. We're leading by example, collaborating with suppliers and tenants to drive decarbonization and embracing sustainable innovation. We invite our peers to join us in this journey, fostering collaboration to reshape the future of real estate." As Mr. Adriel Chan, Hang Lung Properties Chair, noted, "The real estate sector must form a long-term vision together."

The 28 page document sets out the most critical actions to reach net zero by 2050, these are summarised below (Table 6.1):

For Scope 3 emissions, Hang Lung's pathway include:

1. Purchased Goods and Services: Embodied carbon from building materials like steel, aluminium and concrete accounts for 71.6% of Scope 3 emissions.
2. Downstream Leased Assets: Tenant electricity consumption contributes 19.1%.
3. Reduction Targets: reduce Scope 3 emissions from purchased goods and services and downstream leased assets by 25% by 2030, relative to a 2020 baseline.
4. Challenges:
    a. Emissions from tenant fit outs and refrigerants are currently overlooked but are significant contributors.
    b. Life cycle emissions from interiors may exceed those from the building's core and shell over time.
5. Strategies:
    a. Collaboration with tenants to benchmark and reduce emissions.
    b. Adoption of low carbon construction materials and adaptive reuse of properties.

Table 6.1 Hang Lung properties summary of actions and environmental impact

	Action	*Environmental impact and targets*
1	Operational energy efficiency	Reduce greenhouse gas emissions across Scope 1 and 2 by property benchmarking, building optimisation and AI technology
2	Renewable energy procurement	Aim to significantly cut emissions from purchased energy (Scope 2) and downstream leased assets (Scope 3), especially with power purchase agreements (PPAs) in Mainland China
3	Material efficiency in construction	Use technology to optimise construction materials, reduce overbuilding and aim to cut related CO2 emissions by as much as 30%
4	Low carbon materials	Reduce embodied carbon with low carbon steel and bricks
5	Supply chain collaboration	Focus on steel, concrete and aluminium to source low carbon options
6	Adaptive reuse	Minimise embodied carbon from redevelopment by exploring options to repurpose existing buildings
7	Life cycle refrigerant management	Management practices include leakage detection, condition monitoring, and end-of-life disposal which together can substantially reduce emissions
8	Tenant collaboration	Working with tenants to adopt energy efficiency measures and offer tools for energy benchmarking and reductions

(Hang Lung Properties, 2025)

c. Enhanced life cycle refrigerant management practices (Hang Lung Properties, 2025).

Of particular note are the efforts made to reduce the environmental impact of tenant interior fit outs which they note can exceed the building shell and core. The key focus of tenant collaboration is on embodied carbon benchmarking, material reuse and recycling, energy efficiency and taking a role in managing tenant emissions.

Analysis can provide a return on investment both in terms of any financial performance and reduction in GHG emissions to inform decision-making. All the above materials listed and within the Scope 3 boundary are in the hard to abate category, with cement, steel, aluminium and glass at the top of the list both in terms of CO2e and quantity used.

Sustainable Design

A first, and easy to overlook, step in the supply chain is simply reducing the volume of materials used in the designed specification. "Over-engineering" occurs when designers err on the side of caution in managing risk to shield the possibility of damage, often due to a lack of available data or building codes that require additional capacity in case of future need. The result is additional materials in construction and excess capacity in building systems, such as air conditioning, with the resulting impact on energy efficiency, sustainability performance and user comfort (Hill, 2022).

Sustainable design seeks to address over engineering as well as other issues in development projects, and may be active or passive. Active design is based on the electrical and mechanical systems to drive energy efficiency, improve indoor air quality and reduce water consumption while delivering comfortable conditions. Passive design uses the building features itself to reduce energy consumption and materials used. This might include ventilation, shading and facade, building orientation and passive cooling (Architectural Services Department, Government of Hong Kong SAR, 2024).

Choice of materials can help incorporate recycled materials and reduce pollution while prefabrication and modular building can reduce waste while speeding up construction. Design for durability, adaptability and circularity helps extend the life of materials and their reuse. This is an area where policy and building codes can incentivise producers to reduce waste and provide infrastructure and investment to support circularity.

According to a McKinsey & Company publication "Net-zero steel in construction: The way forward", concrete and steel account for around 60% of embodied carbon in a mid-rise (8 storey) commercial building. The means to decarbonise such a building include:

1. Material efficient design practices.
2. Low CO2 steel (abatement potential of ~ 8%).

3. Zero-emissions concrete (abatement potential of ~ 25%).
4. Use alternative structural materials such as cross laminated timber (abatement potential of 20%).
5. Electrification of construction machinery (Assunção et al., 2022).

Additional items are design for disassembly and modular construction. Modular construction has been around for a while but is gaining more and more traction as a means to lower cost and carbon. Some markets like Hong Kong have seen a significant government push for modular construction in public housing, the issue around delivery in narrow streets has been overcome with the use of technology. Design for disassembly helps support circularity, for interiors this would mean using screws instead of glue for example. As highlighted by Barbhuiya et al (2024), the pursuit of alternative materials for decarbonisation drives sustainable construction practices, innovation as well as giving suppliers that embrace decarbonisation a competitive edge (Barbhuiya et al., 2024).

The number of variables in production seems to determine the number of options, or levers, for decarbonisation. Cement has numerous variables both in existing production and options for decarbonisation. Steel and aluminium are broadly similar with key technological breakthroughs still needed to achieve optimum changes in production and cost issues related to upgrading existing projects.

Low Carbon Cement and Concrete

Cement is the binder used in concrete and is made by firing limestone at high temperatures. Since concrete is the cement containing material used in construction it's reasonable to consider the material in use, concrete, rather than the most carbon intensive ingredient, cement. Practically speaking the two are inseparable in the context of this analysis.

The levels or urbanisation expected in Asia and Africa are forecast to result in demand for cement in 2050 to be 12–23% higher than 2020 levels (IEA, 2018), making the need for scaleable sustainable solutions urgent. According to more recent forecasts in a February 2025 article in Global Cement Magazine, demand patterns are more nuanced. There are regional variations in demand for cement. Developed markets of Europe, North America and Oceania are largely stable. Demand from China is falling and expected to stabilise, middle income nations of the Middle-East, North Africa, Latin America, CIS and Southeast Asia are expected to experience steady growth while South Asia and sub-Saharan Africa are expected to see demand double from 2024 to 2050 (Riley, 2025). Overall demand may actually fall as a result of industry-wide action.

Possibly the largest range of innovative options available in any of the hard to abate building materials exist in the cement industry. Starting with not using it at all by using alternative clinker free materials, reducing waste and sustainable design. The University College of Estate Management (UCEM) has identified 12 alternatives to traditional concrete:

1. Green concrete that uses circular economy principles by including waste materials as a cement replacement. This uses less materials and energy and can be more durable. Waste materials include power plant and mining residue materials.
2. Alternative aggregates used in concrete include a variety of waste materials to reduce the quantity of new materials and energy.
3. Ashcrete is a green concrete that uses fly ash from coal power stations. An additional benefit is the greater strength but may take longer to set.
4. Micro silica is a fine powder by-product of the silicon metal industry and could replace 7 to 12% of cement. It may have strength benefits but is harder to work.
5. Blast furnace slag from steel production can be ground into ground granulated blast furnace slag (GGBFS). It is highly workable but may have lower strength and longer setting times.
6. Concrete debris makes use of waste materials from demolition to give a reduced CO2 footprint to the quick to produce new material.
7. Grasscrete incorporates grass and plants into walkways to reduce the amount of concrete needed while increasing natural drainage capacity.
8. Recycled plastic can be used to replace sand and other aggregates but because it doesn't mix so well, the weaker material may not be suitable for all uses.
9. Mycelium derived from fungi is a lighter material and naturally fire resistant but lacks the strength needed in many traditional concreted uses.
10. Hempcrete is made from the fast and easy to grow plant offering a low weight alternative with good thermal properties.

11. Self-healing concrete can help double the lifespan of concrete by using adhesive materials to create bacterial reactions and repair cracks.
12. Bamboo is another fast-growing plant that is strong, flexible and resistant to corrosion and suitable for use in flooring slabs and roofing (UCEM, 2024a, 2024b).

The Massachusetts Institute of Technology Concrete Sustainability Hub (MIT CSHub) seeks to realise a carbon neutral concrete industry. To do this they have started with a comprehensive and verifiable system to consider total impact of the emissions from production, transportation, use within structures and end of life. With a carbon accounting methodology in place, the consideration can then be given to various carbon neutral concrete products, the possibility for CO2 sequestration in concrete during production, circularity in the use of concrete waste, application of sustainable design to improve material efficiency, optimisation of life cycle emissions and development of multifunctional concrete technologies including renewable energy storage (MIT CSHub, 2024).

Additional cement decarbonisation options include the use of calcined clay-based cements as a substitute for clinker. Lower energy production results in belite-based cement products. This can also be combined with improving kiln efficiency by retrofitting old kilns. Other alternatives include magnesium-based cement and alkali-activated binders (Barbhuiya et al., 2024).

While the foregoing offers some insights from academia, both the World Cement Association (WCA) and Global Cement and Concrete Association (GCCA) support full decarbonisation of the industry. In practice this means sharing best practices to manage emissions, promotion of decarbonisation and benchmarking, collaboration and research to share green finance information and other resources, advocating for policy frameworks to reduce CO2 in cement, support energy efficiency, alternative fuel use and other measures to reduce emissions and work with stakeholders in upstream and downstream value chains to minimise carbon footprints (WCA, 2021).

The GCCA Net Zero Roadmap sets out a number of emission reduction measures (Table 6.2).

While there are products described as "green cement" there is no industry standard for what that means and the products are relatively scarce in the market due to the lack of scaleable solutions. Decarbonisation of cement requires a combination of innovative production processes, replacing traditional cement with alternatives and carbon

Table 6.2 Summary of GCCA emission reduction measures

1	Clinker production	Decarbonated raw materials, energy efficiency, kiln electrification or use of hydrogen and use of waste materials	2% reduction in total emissions
2	Clinker substitution/ savings in cement and binders	Use of fly ash, ground limestone, etc	0.5% total reduction in overall CO2 emissions
3	Production efficiency	Scaled up industrial production in fragmented markets	5 to 14% decrease in demand for binder by 2030 and 2050, respectively
4	Carbon capture and storage	CO2 stored within the cement and concrete by injection into wet concrete, curing hardened products and manufacturing aggregates from waste products	
5	Decarbonisation of electricity	Electricity used in production to be zero-carbon by 2050	100% reduction
6	Recarbonation	A natural process and long considered in engineering	Up to 20% is considered a conservative value
7	Sustainable design	Ensure CO2 reduction measures are incorporated into design parameters Includes design of floor slab, column spacing and cement strength	7 and 22% reduction in CO2 emissions by 2030 and 2050, respectively

(GCCA, 2024)

capture according to Amr Nader, CEO of A3&Co in Dubai (Pasquali, 2024). Selected green cement products include:

1. Green Valley Biochar.
2. CarbonCure.
3. Partanna.
4. Hoffman Green (Pasquali, 2024).

Having an industry standard will help. CarbonStar® is a user-friendly, binational, certification standard to enable comparison on embodied carbon in concrete. Jointly developed by the Standards Council of Canada and the American National Standards Institute, it establishes a standard methodology to quantify the carbon intensity of concrete regardless of the materials or production methods used. This is part of "the essential effort of decarbonising concrete" according to Sarah Saltzer, "Managing Director of the Stanford Center for Carbon Storage" (WCA, 2024).

Producers also have a key role to play. "Over the past 30 years the industry has cut per-tonne emissions by about 25%" (Riley, 2025). Major producers with decarbonisation plans include Lafarge, Cemex, Cimat, Arabian and Conch but the use of green cement is not widespread. One showcase example is the use of 99.2% clinker free cement used in a show case 30 cu.m concrete pavement used at Holcim's Seattle Plant in 2024. Holcim is a building materials producer and "global leader in innovative and sustainable building solutions" (Holcim, n.d.).

Decarbonising cement and implementing requirements for green cement into the supply chain requires multiple layers in the process to be aligned. This is not an easy task. Ecocem is Europe's leading provider of high performance, low carbon cement which is manufactured using ACT technology offering the potential to reduce global emissions by 4% by 2035 (Ecocem, n.d.). According to Ecocem, their technology uses clinker substitution to reduce the amount of clinker by 70% and cutting the carbon footprint to around 200 kg CO2/tonne compared to the average of 600 kg.

Other technologies include sequestration, or capturing, the CO2 during the production process. This is expensive, needs additional energy inputs, multi-stakeholder co-ordination and requires advance investment based on variable demand which is subject to project delays (Riley, 2025). Avoiding the production-related carbon in the first place is a better option, but there are other cost and performance factors to consider including strength, setting time, transportation requirements that will determine the type of product suitable for any given project. A significant opportunity exists for non-clinker binders which can be used in "non-structural and sub-surface applications" which account for around 50% of all concrete used (Riley, 2025).

With variations between each cement plant including energy efficiency, type of fuel, age and so on, the solutions in each case vary. A contextual, plant specific low carbon pathway needs to be prepared. Taking the most aggressive low carbon strategies could mitigate most of the emissions increase between 2020 and 2050 (Cheng et al., 2023). In the case of One Taikoo Place, a high rise office building in Quarry Bay, Hong Kong, the use of 75% Ground Granulated Blast Furnace Slag (GGBS) concrete would have resulted in a 21% reduction in carbon emissions (Gan et al., 2023).

The WCA is now forecasting demand for cement, and therefore clinker, to fall. Allowing for wider adoption of clinker free binders and other alternative materials, demand for clinker could fall by as much as half by 2050. At the same time, moving away from coal will reduce production emissions. Overall, the WCA anticipates a fall of more than 50% in cement production CO2 emissions between 2024 and 2050 (Riley, 2025).

Decarbonising Steel Production

According to the World Economic Forum, steel production is a top priority for decarbonisation because it is the top "carbon-emitting manufacturing sector", responsible for 7% of all anthropogenic GHG emissions. It has a knock on effect on a range of other sectors, including the built environment. Any carbon reduction in steel making would help reduce the emissions from all the downstream industries that use it (Stamm & Naujok, 2023). With global steel production estimated to increase by 33% by 2050 (IEF, 2024), decarbonising production is an urgent task.

A first step for making steel making greener is to retrofit conventional blast furnaces to make them cleaner and more energy efficient. Carbon capture utilisation and storage (CCUS) is another emerging technology but yet to be widely accepted. Having initially had some support, it now seems less likely to play a major role and there are no projects at scale that have shown the technology to be commercially viable (Nicholas & Basirat, 2024).

Green hydrogen and renewable energy electric arc furnaces (EAFs) are gaining traction and a number of hydrogen-based EAF projects are under development despite the high cost of hydrogen. Currently the combined use of renewable energy and increased use of recycled steel are the economically viable means of producing low carbon steel. The problem with EAF is that they only work with scrap metal or steel pellets that have already been made using carbon intensive methods (IEF, 2024) and the problem with green hydrogen is the cost. This is where the role of carbon pricing helps. With a carbon price of US$50 per tonne the cost of hydrogen needs to fall to US$1.60 per kg (Hydrogen Council, 2020), compared to US$5 in 2022.

The Mission Possible Partnership steel sector transition strategy offers 10 key actions and insights to net zero steel by 2050. These are summarised as follows:

1. Implement a full range of existing and new technologies.
2. Achieving the 2050 target requires a start in the early 2020s.
3. The progress made by 2030 will determine the technology being used in 2050 due to the lifespan of production facilities.
4. Increased use of scrap is critical, efficient use of materials and the use of gas-based direct reduced-iron steelmaking.
5. Tailoring technology appropriate for different locations according to available resources, existing production, policies and demand factors.
6. Residual emissions will always remain and will need to be addressed.
7. Investment to transform the industry by commercialising technologies.
8. Policies to manage issues related to the price premium for low carbon steel.
9. Change in energy inputs from coal to renewable electricity, green hydrogen and natural gas.
10. Support mechanisms to establish the first generation of low carbon steel production projects (Mission Possible Partnership, 2022).

In the case study of One Taikoo Place, the potential use of 100% recycled EAF steel rebar would result in a 64% decrease in the emissions factor of that product or 34% less for the whole design (Gan et al., 2023).

Aluminium

Aluminium is second only to steel in metal used for all functions globally and is a critical material in grid infrastructure, wind turbines, solar power generation and electric vehicles (Wells & Ahmed, 2023). In buildings, aluminium is used in window frames, curtain walls, doors and other uses where lightweight properties and corrosion resistance are advantageous. It tends to be more energy intensive to make than steel, resulting in higher emissions 17.5 kg CO2e/kg vs 3.9 (He et al., 2020) and accounts for 3% of global direct industrial CO2 emissions (IEA, 2023).

According to the European Commission, "Decarbonisation Options for the Aluminium Industry" decarbonisation is possible if technological and economic hurdles can be overcome. Similar to steel production, CCUS is a possible decarbonisation pathway but is less efficient than in steel due to lower CO2 concentrations in the smelting process. Other options include hydrogen and the electrification of smelting and recycling. Inert anodes, made from ceramics or alloys, instead of carbon are the key technological breakthrough that eliminates GHG emissions with oxygen produced instead of CO2 and other gases. So far, the technology remains out of the reach of commercial application but Elysis is a Canadian firm, a partnership between Alcoa and Rio Tinto, using the technology to eliminate all GHG emissions from the aluminium smelting process (Elysis, n.d.). Additional benefits include safer production processes and lower operating costs.

In common with decarbonising production of cement and steel, carbon pricing and financial levers are key mechanisms for levelling the playing field to make the new technology economically viable. Recycling is the quickest and easiest option; the use of scrap in secondary production offers energy savings of 95% compared to primary production (IAI, n.d.).

Flat Glass

Flat glass used in windows and building facades accounts for 33% of total glass industry-related emissions compared to 47% for container glass (Zier et al., 2021). Improvements in the performance of facades for thermal insulation and control of thermal gain have resulted in glass being a key building material and a critical part of built environment decarbonisation.

75% of flat glass manufacturing emissions are from fuel (natural gas) combustion and 25% are process emissions (Glass for Europe, 2020). Thus the decarbonisation efforts are largely energy related with recycling as a further option, albeit not a simple one.

Zier et al (2021) detail the various decarbonisation efforts in the glass industry, while Glass for Europe present the leading initiatives from the flat glass industry in the EU.

Glass decarbonisation options fall into categories of:

1. Waste heat recovery.
    a. Combustion gases and oxygen preheating.
    b. Batch and cullet preheating.
    c. Thermo-chemical heat recovery.
    d. Heat-to-Power.
    e. Building/district heating.
2. Fuel substitution.
    a. Electrification.
    b. Biogas.
    c. Solid biomass and gasification
    d. Hydrogen.
    e. Synthetic methane.
    f. Microwaves.
    g. Plasma.
3. Process intensification.
    a. Furnace design and operation.
    b. Submerged combustion melting.
    c. Selective batching.
    d. Pelletisation.
    e. Fluxing agents.
    f. Advanced process control.
    g. Batch wetting.
4. CO2 pricing (Zier et al., 2021).

While academia provides the comprehensive and thorough list of all the options in a complex industrial process, the industry association publication offers what might be a more practical application, albeit as a document to promote the decarbonisation efforts of the industry in the EU. These are presented as:

1. Furnace design, construction and operation—estimated CO2 savings of around 25% since 1990.
2. Raw materials and recycled content—estimated 6% CO2 reductions.
3. Energy mix—potential CO2 reduction of 30% (Glass for Europe, 2020).

The consensus is that energy substitution has the greatest potential decarbonisation impact with hydrogen considered a potential commercially viable alternative to natural gas. Up to 20% hydrogen can be used without retrofitting the furnaces (Glass for Europe, 2020). Biofuel has also been used to produce glass with 80% lower CO2 emissions than natural gas (Sanders, 2024). If the required technological advances are made, electric melting furnaces have the highest potential to decarbonise and can be combined with other options to deliver an optimised approach.

Increasing recycled content very much depends on supporting policies to recover clean glass from demolition sites and transport it at reasonable cost for processing. The quality standards for glass used in building facades are high and this limits the amount of recycled content, known in the industry as cullet, that can be used.

Building Materials Procurement

The procurement process is a critical part of any construction project and in the case of concrete, for example, may not be controlled by the developer or general contractor but a subcontractor (Pasquali, 2024). The use of green procurement principles can support the reduction of embodied carbon in development of new buildings and refurbishment projects and means considering "environmental concerns into major purchasing strategies, policies, directives and practices" (Ling, 2020) to help reduce the overall environmental impact of the project.

One key recommendation from (Gan et al., 2023) was to require the relevant carbon data from the manufacturer as part of the procurement process. Another key point was to focus on the rebar, structural steel and concrete that account for 89.2% of production emissions at One Taikoo Place as well as their transportation, 5.5%, and construction energy, 3.6%. At less than 1%, glass may be set aside for the time being (Gan et al., 2023).

Procurement is one of the responsibilities of the project manager and includes:

1. Definition of the project requirements.
2. Tendering and supplier selection.
3. Acquisition of supplies and equipment.
4. Contract, cost and completion management.
5. Budget, timeline and quality conflict resolution (Lieberson, 2024).

Waris et al., (2018) note the environmental benefits through use of green materials, water management and reduced energy consumption. In addition to social benefits associated with reduced accidents in construction and a healthier working environment, economic benefits through efficient procurement and recycling and political benefits by alignment with government environmental plans. At the same time, barriers include lack of corporate commitment, cost premiums, knowledge gaps, lack of guidance and information, a business as usual approach and few supporting regulations. Additional barriers include inconsistent policies, vague definitions, separation of operational and capital budgets, a short-term view, perception of higher costs, lack of industry collaboration and insufficient research and development. The role of regulators is considered the most important in unlocking these barriers (Sourani & Sohail, 2011).

The Role of Partnerships

A few partnerships in the production processes have been mentioned already. Sustainable Development Goal 17 is strengthen the means of implementation and revitalise the Global Partnerships for Sustainable Development (UN, n.d.). This includes topics such as multi-stakeholder partnerships, capacity building, finance, science and technology which are all relevant to this chapter. As stated by Cate Harris, Group Head of Sustainability and Lendlease Foundation, on their website "No one organisation can solve the climate crisis, but together we can catalyse industry collaborations to encourage investment to test, scale and drive the necessary transformation at pace".

In Hong Kong and mainland China, a number of listed property developers have established partnership arrangements. These include upstream partnerships with material suppliers and downstream partnerships with tenants. Such partnerships enable organisations to stretch further and aim higher, John Haffner is Deputy Director, Sustainability at Hang Lung Properties commented "our corporate 2030 steel target through SteelZero is ambitious to say the least; we are stretching ourselves to try to help build a demand signal and accelerate progress in steel decarbonization. And we're glad that we've been able to make initial steps forward with our lower carbon emissions steel procurement at our Plaza 66 extension project, and with more good news to follow soon on our Westlake 66 project." A summary of a selection of partnerships is provided below (Table 6.3):

The range of partnerships above is both inspiring and revealing. Upstream partnerships are apparently fewer, probably more technically complex to execute but potentially transformative. While downstream ones demonstrate the opportunity to develop closer tenant relations as a means to improve tenant satisfaction, are critical in reducing tenant fit our embodied carbon and improve sustainability ratings for landlords and tenants with some reputational benefits too. The table above shows the extent to which both upstream and downstream partnerships have evolved to deliver impactful change, at the same time, the potential for wider industry adoption of the partnership model is revealed. No doubt more will be announced in the coming years as all parties begin to see the benefits.

Circular Business Model for the Built Environment

The circular business model takes the linear "business-as-usual" model (take resources, make stuff, throw them away) and creates products that can be repurposed, reused or recycled. This can help improve efficiency, reduce waste and limit environmental impact (UCEM, 2024a, 2024b).

Table 6.3 Summary of selected published sustainability partnerships in Mainland China and Hong Kong SAR

Developer	*Partners*	*Description*	*Remarks*
Hang Lung Properties	LVMH	Established in October 2022 to collaborate on sustainability initiatives across mainland China. Includes 26 LVMH brands and over 27,000 sq.m of retail space	Real Estate and Climate Change Forum held in 2022 followed by a Common Charter: Joining Forces to Accelerate Change including 20 actions under the headings of resource management, climate resilience, people and well-being and sustainability governance. The Charter includes additional external partnerships
Hang Lung Properties, Swire Properties and Shui On Land	China Iron and Steel Association, Urban Land Institute Greenprint and World Steel Association	Collaboration on Low Carbon Emissions Steel for Real Estate in China. Comprises a pledge to accelerate progress in reducing emissions from steel in real estate	Includes procurement, guidelines, disclosure and transparency and collaboration with peers and stakeholders. Results in 3317 tonnes CO2 emissions reduction for Shui On projects in 2024 and 35% reduction in steel-related CO2 emissions for Hang Lung Plaza 66 Pavilion Extension

(continued)

Table 6.3 (continued)

Developer	*Partners*	*Description*	*Remarks*
Swire Properties	Various office tenants	Green Performance Pledge a "landlord-tenant partnership and performance based initiative that aims to promote joint action towards shared sustainability goals."	By late July 2024 "54% of the occupied lettable space" in the Hong Kong portfolio had joined the pledge and had collectively achieved a 3.8% reduction in energy use intensity (EUI) and a 10% increase in waste diversion rates
Swire Properties	LVMH	Strategic Sustainability Partnership incorporating GPP, Green Retail Partnership, Green Kitchen Initiative and Eco-design Checklist	Covers around 50,000 sq.m of office, retail and F&B premises
Hongkong Land	Various office tenants	Tenant Sustainability Partnership Programme to collaborate, share, recognise and reward. Includes environmental and social initiatives	Focus areas include green fit outs, green operations, tenant satisfaction surveys, diversity and inclusion, volunteering, ESG events and donations

(continued)

Table 6.3 (continued)

Developer	*Partners*	*Description*	*Remarks*
Nan Fung Group	Tenants, employees and community partners	"Net Positive Lease" to offer incentives and benefits to mobilise tenants to achieve sustainability goals with the landlord. Includes data sharing and analytics and social impact assessment	First to offer management fee incentive in Hong Kong to participants. By March 2024, 21% of tenants had committed and expansion to properties in the UK, US and China is planned
Chinachem	China Light and Power (CLP)	Collaboration for energy efficiency, electrification and decarbonisation	Original partnership dates from 2019 resulting in around 14 GWh of energy savings in Nina Tower by late 2023. New MoU signed in November 2023 implements an energy management system to deliver precision cooling to Nina Mall and extends energy audit collaboration to cold storage and logistics including a green data centre. Other initiatives include battery storage on construction sites, sustainability-linked loans and promoting energy efficiency to tenants

(continued)

Table 6.3 (continued)

Developer	*Partners*	*Description*	*Remarks*
Henderson Land	Various tenants and their employees in The Henderson	Incentive-based tenant and employee reward programme and data-driven platform for tenants to track ESG performance	Includes a smart APP for tenants and employees to earn participation-based rewards Digital Twin powered ESG data for 12 ESG metrics to help tenants manage sustainability performance

(Hang Lung Properties & LVMH, 2022; Swire Properties, 2024a, 2024b, 2024c; Hang Lung Properties, 2024; Shui On Land, 2024; Swire Properties, 2024a, 2024b, 2024c; Hong Kong Land, 2023; Nan Fung Group, 2023; CLP, 2023; Swire Properties, 2024a, 2024b, 2024c; Henderson Land, 2022).

As explored above, the inclusion of recycled materials into the production process is key to decarbonising the hard to abate materials. Steel and aluminium are readily recyclable, concrete and glass can incorporate recycled content in their manufacturing process to a lesser extent.

Making it happen requires collaboration and partnerships. The previous section listed a number of Hong Kong-based companies that are already moving towards these new partnerships in a number of ways. Concepts like circularity and sustainable design are rarely part of a property developer mindset, they have to identify them as a strategic goal, define the possible and work hard to embed the skill sets across the organisation. They need help to understand how to incorporate sustainability into existing structures and see the importance of future proofing, for example, rainwater harvesting may be a totally new idea for many clients that requires upfront investment for a longer term gain. Pockets of misinformation may persist at board level.

Lots of good information is freely shared and publicly available. The Ellen MacArthur Foundation (EMF) is a charity committed to creating a circular economy for materials and products while eliminating waste (Ellen MacArthur Foundation, n.d.). Circular Economy 100 (CE100) is an EMF initiative to establish a platform to bring stakeholders together to help speed up the transition to a circular economy (WGBC, n.d.). This includes circular cities and the built environment and in 2020 Amsterdam became the first city to set a goal of becoming a 100% circular economy by 2050. For the built environment this means embedding circularity into new and existing developments, encouraging innovation and supporting the market for recycled and recyclable materials (Ellen MacArthur Foundation, 2024).

Carra & Magdani, (n.d.) identified enabling factors of design for deconstruction, reassembly and future flexibility. Under information they consider cost, life cycle data, traceability, and ownership. While collaboration in terms of incentives, innovation, term of investment and transparency are enablers. Circular design presents opportunities to extend life of buildings and enjoy a higher residual valuation being easier to upgrade, maintain or recycle. The use of less resource intense materials may have the potential to be reused or recycled and there may be new uses for building components. Challenges include higher upfront costs, slow market acceptance and technological and operational uncertainties.

The WGBC Circular Environment Playbook details the stakeholders, the key themes, the levers for change and a call to action with a range of case studies. What stands out are things like material databases to keep the full inventory of what has gone into the building, the concept of buildings as ‘material banks’ ready for disassembly and reuse and materials cascading through multiple product uses before finally being incinerated for energy. A ‘circular-ready’ checklist for the built environment details all the steps required and responsibilities for building construction and materials, design and retrofit, nature regeneration and levers for change (Jackson et al., 2023).

The theoretical work for circularity in the built environment has all been done. Carra and Magdani call for:

1. Longer term perspectives.
2. Design for disassembly.
3. Innovation.
4. Durability and flexibility balance.

5. Reimagining production and consumption.
6. Collaboration.

While Jackson et al. identify next steps as:

7. Setting out the closed-loop vision.
8. Consistent measurement and indicators.
9. Awareness and upskilling
10. Policy change (Jackson et al., 2023).

Taken together 10 items, all of which can be imagined and achieved. Some of which add value. "Value is added by increasing the ability of assets to respond flexibly to market conditions, increasing asset use, diversifying income streams and maximizing the residual value of a building's materials" (Carra & Magdani, n.d.).

Return on Investment

Costing the transformation required is of course difficult and the market will respond as best it can. One key concept is "abatement cost", that is the cost of intervening to reduce one tonne of greenhouse gas emissions (Hallegatte, 2023). This can help set actionable priorities and estimate return on investments when other inputs are understood, including the cost of doing nothing, actual or potential carbon tax avoidance and so on.

An abridged quote from Stephane Hallegatte, Chief Climate Economist at World Bank Group, reads: "Retrofitting all buildings has a high abatement cost, and we could therefore be tempted to begin doing so in 2040, for example. But we need to reach carbon neutrality by 2050, and retrofitting all buildings over a 10-year period is completely unrealistic." ... "Even if the cost is higher than for other options, we cannot wait until 2040 to begin. In this case, what determines the most economically efficient sequencing is not the marginal cost for each individual building, but the cost for all buildings. Going too quickly is expensive, as there are not enough resources or qualified workers. Going too slowly is also expensive, as it means economies of scale cannot be achieved (for example in training). Knowing that buildings need to be renovated by 2050, the challenge is to determine the best strategy to do so at low cost. Comparing the costs per tonne in the building sector to the costs per tonne in the forestry sector (or any other sector) is pointless, since we know that the work will need to be done in all sectors" (Hallegatte, 2023).

Returning to the McKinsey Sustainability report "Building Value by Decarbonising the Built Environment" we learn that nearly half of all built environment-related emissions can be abated with zero net cost. That is the savings in operational terms justify the upfront cost premium. For embodied carbon, McKinsey identify key actions:

1. Green material production.
    a. Low carbon cement by retrofitting cement kilns.
    b. Scale up low carbon concrete production by incorporating more recycled content and carbon curing.
    c. Green steel production with use of scrap, adoption of green hydrogen furnaces and carbon capture.
    d. Low carbon insulation by adopting fast-growing natural fibres such as hemp.
    e. Adoption of engineered wood to replace steel and concrete.
2. Cost-effective green designs.
    a. Upskill professionals to prioritise green solutions and educate their clients.
3. Electrification of construction sites.
    a. Prioritise grid connection.
    b. On-site batteries to replace diesel generators.
    c. Electric construction equipment, e.g. crawler cranes.
4. Maximise off-site construction.
    a. Reduce waste and save costs compared to on-site construction.
    b. Standardise design and use modular construction to drive efficiency.
5. Green financing.
    a. Overcome additional up-front costs.
    b. Specialist green retrofit finance.

c. Identify specialist insurers for construction using innovative green designs, materials and construction (Boland et al, 2023).

Development of clusters of green technology industries will enable collaboration, integration and scaling up. Leaving Stephane Hallegatte with the last word on this "We need to implement integrated strategies, which are the only ones capable of organising—in all sectors simultaneously—an economic, technological and social transition towards carbon neutrality." For embodied carbon in the built environment, this chapter has certainly shown this to be the case.

Summary

The climate crisis and race to net zero are based on complex science; it's easier to carry on with business as usual than it is to make the effort to try to understand. Building and construction professionals are smart people, but perhaps not hard wired to work on climate-related non-financial metrics. Just as they accept the science of aviation every time they board a plane, they should also learn to accept the science of climate change and the technology to reduce emissions with the emerging technology without feeling the need to understand all the details.

The subject matter in sustainability and buildings is new and not covered in qualifications achieved more than 5 or 10 years ago. It is not surprising that business leaders in the twilight of their property development careers are happy to dismiss the complexity of negotiations with loyal suppliers for procurement of low carbon cement or green steel.

The foregoing sections of this chapter aim to cut through the complexity and empower those with the authority to join the movement towards collective action across the built environment value chain. The environmental problems are understood and the solutions are available or within our grasp.

The key element is at the top of the value chain; the property developers have to commit to buying low carbon materials at scale to enable the manufacturers to talk to their bankers and make the investment in the production upgrades.

References

Architectural Services Department, Government of Hong Kong SAR. (2024). *On the state of sustainable built environment*. Retrieved November, 2024, from https://www.archsd.gov.hk/archsd/html/report2024/en/index.html

Barbhuiya, S., Kanavaris, F., Das, B. B., & Idrees, M. (2024, June 1). *Decarbonising cement and concrete production: Strategies, challenges and pathways for sustainable development*. Retrieved December, 2024, from https://www.sciencedirect.com/science/article/pii/S2352710224004297

Assunção, P., Graham, A., Imhorst, F., Koester, I., Kühl, C., Sullivan, R., Ulanov, A., D'Avolio, E., Burns, T., & Boland, B. (2022, April 28). *Net-zero steel in building and construction: The way forward*. McKinsey. Retrieved August 26, 2025, from https://www.mckinsey.com/capabilities/sustainability/our-insights/net-zero-steel-in-building-and-construction-the-way-forward

Boland, B., Lekhwani, S., Reiter, S., & Sjödin, E. (2023, June). *Building value by decarbonizing the built environment*. Retrieved August 21, 2025, from https://www.mckinsey.com/industries/engineering-construction-and-building-materials/our-insights/building-value-by-decarbonizing-the-built-environment

Carbon Trust. (n.d.). *What are Scope 3 emissions and why do they matter?* Carbon Trust. Retrieved December 18, 2024, from https://www.carbontrust.com/our-work-and-impact/guides-reports-and-tools/what-are-scope-3-emissions-and-why-do-they-matter

Carra, G., & Magdani, N. (n.d.). *Circular business models for the built environment*. Retrieved December, 2024, from https://emf.thirdlight.com/file/24/P-UFl1YP4ZV3AhP-IWdPrRtzn9/Circular%20Business%20Model%20For%20the%20Built%20Environment.pdf

Cheng, D., Reiner, D., Yang, F., Cui, C., Meng, J., Shan, Y., Liu, Y., Tao, S., & Guan, D. (2023, December 11). *Projecting future carbon emissions from cement production in developing countries*. Nature Communications. https://www.nature.com/articles/s41467-023-43660-x

CLP. (2023, November 9). *Chinachem Group and CLP Power enter new sustainability partnership to explore more decarbonisation and electrification opportunities*. Retrieved December, 2024, from https://www.clpgroup.com/content/dam/clp-group/channels/media/document/2023/20231109_en.pdf.coredownload.pdf

Ecocem. (n.d.). *Raising ambitions, reducing emissions*. Ecocem - The future. Remixed. Retrieved June 25, 2025, from https://www.ecocemglobal.com/?

Ellen MacArthur Foundation. (n.d.). *What we do | Ellen Macarthur Foundation*. Ellen MacArthur Foundation. Retrieved December 18, 2024, from https://www.ellenmacarthurfoundation.org/about-us/what-we-do

Ellen MacArthur Foundation. (2024, January 10). *Aiming for 100% circularity: Amsterdam*. Retrieved December, 2024, from https://www.ellenmacarthurfoundation.org/circular-examples/shaping-a-sharing-economy-amsterdam

Elysis. (n.d.). *What is ELYSIS?* ELYSIS. Retrieved June 25, 2025, from https://elysis.com/en/what-is-elysis

Gan, J., Li, K., Li, X., Mok, E., Ho, P., Law, J., Kwok, R., & Yau, R. (2023). Parametric BIM-based lifecycle performance prediction and optimisation for residential buildings using alternative materials and designs. *Buildings, 13*(4), 904. https://doi.org/10.3390/buildings13040904

Glass for Europe. (2020). *Flat Glass in climate-neutral Europe*. Glass for Europe. Retrieved May 31, 2024, from https://glassforeurope.com/wp-content/uploads/2020/01/flat-glass-climate-neutral-europe.pdf

Global Cement and Concrete Association. (2024). *Getting to Net Zero*. Retrieved December, 2024, from https://gccassociation.org/concretefuture/getting-to-net-zero/

Hallegatte, S. (2023, April 4). *Proper use of the abatement cost to steer the transition*. Institute for Climate Economics. Retrieved November, 2024, from https://www.i4ce.org/en/proper-use-abatment-cost-streer-transition-climate

Hang Lung Properties. (2025, March). *Our journey to Net Zero*. Retrieved April 10, 2025, from https://www.hanglung.com/getmedia/c22fa7da-a67b-49e3-a09b-f70f4b6b39c8/Our-Journey-to-Net-Zero-EN.pdf

Hang Lung Properties. (2024, December). *p u n 100% lc e s*. Retrieved December, 2024, from https://www.linkedin.com/posts/hanglung_low-carbon-emissions-steel-activity-7267714318232621057-c23j/

Hang Lung Properties & LVMH. (2022, October 25). *Hang Lung Properties and LVMH Group launch a new model of sustainability partnership*. Retrieved December, 2024, from https://www.hanglung.com/getmedia/2c2d5bcb-3106-469f-9b29-0a44f0d3809f/20221025_Press-Release-Final2_EN_1.pdf

He, X., Kim, H. C., Wallington, T. J., Zhang, S., Shen, W., De Kleine, R., Keoleian, G. A., Ma, R., Zheng, Y., Zhou, B., & Wu, Y. (2020, January). Cradle-to-gate greenhouse gas (GHG) burdens for aluminum and steel production and cradle-to-grave GHG benefits of vehicle lightweighting in China. *Resources, Conservation and Recycling*, *152*. https://doi.org/10.1016/j.resconrec.2019.104497

Henderson Land. (2022, May 19). *Henderson Land Launches Industry-first ESG Partnership Programme at "The Henderson" — a Tri-party Collaboration Empowering Tenants and Their Employees to Strive towards Sustainability Goals*. Retrieved December, 2024, from https://www.hld.com/en/our-news/press-releases/henderson-land-launches-industry-first-esg-partnership-programme-at-the-henderson-a-tri-party-collaboration-empowering-tenants-and-their-employees-to-strive-towards-sustain

Hill, A. (2022, February 21). *Over-engineered buildings obstruct efficiency.* Retrieved November, 2024, from https://www.designingbuildings.co.uk/wiki/Over-engineered_buildings_obstruct_efficiency

Hill, L. (2019, August 2). *Float glass vs. flat glass vs. plate glass vs. sheet glass.* Glass.com. Retrieved May 15, 2022, from https://info.glass.com/float-flat-plate-sheet-glass/

Hines. (n.d.). *About Hines.* Hines: Global Real Estate Investment Manager. Retrieved December 18, 2024, from https://www.hines.com/

Hines. (2022, April). *Embodied carbon reduction guide.* Retrieved December, 2024, from https://www.hines.com/embodied-carbon-reduction-guide

Holcim. (n.d.). *About us.* Holcim. Retrieved June 25, 2025, from https://www.holcim.com/who-we-are

Hongkong Land. (2023, June). *Tenant sustainability partnership programme.* Retrieved December, 2024, from https://hklandblob.blob.core.windows.net/assets/sustainability/tenant-engagement/hongkongland-tenant-sustainability-partnership-programme.pdf

HSBC. (2025). *Decarbonising the built environment* [Persepectives from mainland China and Hong Kong].

Hydrogen Council. (2020, January 20). *Path to hydrogen competitiveness.* Retrieved November, 2024, from https://hydrogencouncil.com/wp-content/uploads/2020/01/Path-to-Hydrogen-Competitiveness_Full-Study-1.pdf

IAI. (n.d.). *Aluminium recycling saves 95% of the energy needed for primary aluminium production.* International Aluminium Institute. Retrieved June 25, 2025, from https://international-aluminium.org/landing/aluminium-recycling-saves-95-of-the-energy-needed-for-primary-aluminium-production/

IEA. (2018, April 6). *Technology roadmap - Low-carbon transition in the cement industry – Analysis.* IEA. Retrieved June 25, 2025, from https://www.iea.org/reports/technology-roadmap-low-carbon-transition-in-the-cement-industry

IEA. (2023, July 11). *Aluminium - IEA.* International Energy Agency. Retrieved December 18, 2024, from https://www.iea.org/energy-system/industry/aluminium

International Energy Forum. (2024, January 23). *Cutting the carbon intensity of steel using hydrogen and renewables.* International Energy Forum - IEF. Retrieved December 18, 2024, from https://www.ief.org/news/cutting-the-carbon-intensity-of-steel-using-hydrogen-and-renewables

IMO. (n.d.). *IMO and the sustainable development goals.* International Maritime Organization. Retrieved February 28, 2024, from https://www.imo.org/en/MediaCentre/HotTopics/Pages/SustainableDevelopmentGoals.aspx

Jackson, A., Brady, C., & Owen, C. (2023, May 10). *The circular built environment playbook*. Retrieved December, 2024, from https://worldgbc.org/article/circular-built-environment-playbook/

Lendlease. (n.d.). About Us | Australia. Lendlease. Retrieved December 18, 2024, from https://www.lendlease.com/au/about-us/

Lieberson, M. (2024, January 9). *The construction procurement process, explained*. Retrieved November, 2024, from https://www.quickbase.com/blog/procurement-in-construction-management

Ling, W. P. (2020, July 25). Sustainable procurement for green building. *SIPMM Publications*. https://publication.sipmm.edu.sg/sustainable-procurement-for-green-building/

Mission Possible Partnership. (2022, September). *Making Net-Zero steel possible*. Retrieved November, 2024, from https://3stepsolutions.s3-accelerate.amazonaws.com/assets/custom/010856/downloads/Making-Net-Zero-Steel-possible-steel.pdf

MIT Concrete Sustainability Hub. (2024). *Carbon neutrality*. Retrieved November, 2024, from https://cshub.mit.edu/carbon-neutrality/

Nan Fung Group. (2023, August 8). *Nan Fung Group Launches 'net positive lease'*. Retrieved December, 2024, from https://www.nanfung.com/en/about-us/announcement/nan-fung-group-launches-net-positive-lease-the-first-tenant-engagement-initiative-in-hong-kong-to-offer-management-fee-incentive-nan-fung-collaborates-with-tenants-employ/

Nicholas, S., & Basirat, S. (2024, April 17). *Carbon capture for steel?* IEEFA. Retrieved December 18, 2024, from https://ieefa.org/resources/carbon-capture-steel

Pasquali, V. (2024, October 25). *'Green' cement falling short of meeting climate goals*. Arabian Gulf Business Insight. Retrieved November, 2024, from https://www.agbi.com/analysis/construction/2024/10/green-cement-falling-short-of-meeting-climate-goals/

Riley, I. (2025, February). Cement Demand Forecast 2050. *Global Cement Magazine*, 12–18.

Sanders, H. (2024, September 5). Decarbonizing flat glass. USGlass Magazine. Retrieved November, 2024, from https://usglassmag.com/insights/2024/09/decarbonizing-flat-glass/

Shui On Land. (2024, December). *Shui on land joins pioneering initiative for low-carbon steel in chinese real estate*. Retrieved December, 2024, from https://www.linkedin.com/company/shui-on-land/posts/

Sourani, A., & Sohail, M. (2011, December). Barriers to addressing sustainable construction in public procurement strategies. *Proceedings of the Institute of Civil Engineers: Engineering Sustainability*. https://doi.org/10.1680/ensu.2011.164.4.229

Stamm, H., & Naujok, N. (2023, August 29). *How the steel industry could lead the way in decarbonization*. The World Economic Forum. Retrieved December 18, 2024, from https://www.weforum.org/stories/2023/08/why-steel-can-be-an-unexpected-leader-in-decarbonization/

Swire Properties. (2024, July 29). *Swire properties' "green performance pledge" exceeds 50% target participation rate*. Swire Properties. Retrieved December 18, 2024, from https://www.swireproperties.com/en/media/press-releases/2024/20240729_gpp/

Swire Properties. (2024, November 6). *LVMH and Swire Properties announce strategic sustainability partnership covering retail, office and F&B premises*. Retrieved December, from https://www.swireproperties.com/en/media/press-releases/2024/20241106_lvmh/

Swire Properties. (2024, December). Retrieved December, 2024, from https://www.linkedin.com/posts/swire-properties_swireprops-sd2030-steel-ugcPost-7270706368976769024-CbBY/

UCEM. (2024, January 17). *5 circular business models (and how they can give you a competitive advantage)*. University College of Estate Management. Retrieved December 18, 2024, from https://www.ucem.ac.uk/whats-happening/articles/circular-business-model/

UCEM. (2024, February 12). *12 sustainable alternatives to traditional concrete*. Retrieved December, 2024, from https://www.ucem.ac.uk/whats-happening/articles/concrete-alternatives

UN. (n.d.). *Goal 17 | Department of Economic and Social Affairs*. Sustainable Development Goals. Retrieved December 18, 2024, from https://sdgs.un.org/goals/goal17

Waris, M., Lee, C.-K., Darun, M. R., Ting, N. H., Ullah, M., & Khamidi, M. F. (2018, January). Green procurement in construction industry: A theoretical perspective of enablers and barriers. *MATEC Web of Conferences*. https://doi.org/10.1051/matecconf/201820302012

WBCSD. (2023). *Embodied building emissions can be reduced by 50%*. Retrieved August 12, 2025, from https://www.wbcsd.org/news/embodied-building-emissions-can-be-reduced-by-50-wbcsd-arup-to-tell-property-developers-at-mipim/

WGBC. (n.d.). *The circular economy*. Retrieved December, 2024, from https://worldgbc.org/article/the-circular-economy/

World Cement Association. (2021, December 9). *WCA climate policy*. Climate Policy and Pledges. Retrieved November, 2024, from https://www.worldcementassociation.org/images/download-selector/climate-policy-and-pledges/WCA_Climate_Policy.pdf

Wells, I., & Ahmed, S. (2023, August 10). *The role of inert anodes in aluminum decarbonization*. NRDC. Retrieved June 25, 2025, from https://www.nrdc.org/bio/ian-wells/role-inert-anodes-aluminum-decarbonization

Zier, M., Stenzel, P., Leander, K., & Stolten, D. (2021, May). A review of decarbonization options for the glass industry. *Energy Conversion and Management*, *10*. https://doi.org/10.1016/j.ecmx.2021.100083

CHAPTER 7

Decarbonising Building Operations

"It's Not Easy Bein' Green" Kermit the Frog (written by Joe Raposo).

Building Operations

The previous chapter considered the various means of measuring and reducing embodied carbon while later chapters will discuss the various tools for enabling change, including green building certification and other benchmarking standards. Decarbonising building operations starts with adopting energy management technology to optimise efficiency and removal of all fossil fuels from the building operation supply chain. Building electrification and buying green energy tariffs goes a long way. Other decarbonisation efforts include upgrading chillers to use low Global Warming Potential (GWP) refrigerants, use of heat pumps, transparency of emissions by publication of data, upskilling property management and engineering teams and implementation of green lease agreements and sustainable fit out guides for tenants.

The IEA estimate that air conditioning will account for 37% of the growth in electricity demand between 2018 and 2050. Energy efficiency improvements in building plant and machinery are common measures taken and are already quite well documented among building decarbonisation, they are not the focus of this chapter. The case for electrification, refrigerants, building envelope and tenant fit outs are generally less well

S. Crispin, *Sustainability in the Built Environment*,
https://doi.org/10.1007/978-981-95-6549-8_7

covered and will be covered in more detail. This is a view shared in "Our Journey to Net Zero" which sets out the various scenarios and actions to reduce greenhouse gas emissions by Hang Lung Properties (Hang Lung Properties, 2025). While tenant interior fit outs involve embodied carbon in the materials, these are incurred during the operations stage of the life cycle so are included in this chapter rather than the previous one. The chapter will conclude with green lease agreements which can be used as a tool to focus attention of both landlords and tenants on mutually agreed building sustainability targets, including some of the less well-known topics explored below.

The drivers for change, including policy, regulations and reporting expectations all form part of the overall transition to a low carbon model. There are also financial benefits from more efficient operations. When it comes to implementation, including building operations, it comes down to the various players in the value chain who must each define the problem, set a baseline and boundaries, establish targets and implementation plans, prepare solutions, put everything in place and monitor progress against targets (Cary Chan, 2024).

From Beijing, Sunny Zhang offered some insights into building operations. "Green operations are by no means high-investment "image projects", but a practical path for commercial real estate to reduce costs, increase efficiency, and connect with green finance. BOMA's industry practice data shows that through best practices, such as energy-saving retrofits, preventive maintenance, and low-cost optimization, properties can achieve 10–40% energy savings, directly increasing Net Operating Income (NOI). This direct connection between investment and return is exactly what green financiers are looking for".

She added: "If green operations are regarded as the "engine" of value creation, then the international sustainable operation and management system is the "scale" for value measurement, providing a cross-regional and comparable evaluation framework for green finance. As a globally recognized sustainable management system, the core value of BOMA BEST® Certification lies in transforming scattered green practices into systematic performance indicators, enabling the financial market to accurately identify the sustainable value of assets. By 2025, a number of benchmark projects, including Guangzhou International Finance Center and Shanghai Tower, have obtained financing cost advantages and asset valuation improvements through this system certification".

A telling case study for the confluence of drivers for change and implementation planning is the Empire Building Challenge, a public-private partnership with leading property owners and comprising investment from New York State to prove the case for decarbonisation of tall buildings in cold climates. The case of the Empire State Building in New York is discussed in more detail in Chapter 10.

Building Electrification

Building electrification is a key step to eliminating fossil fuels from buildings. It involves the conversion of all fossil fuel burning plant, machinery and installations in the building to electric only equivalents. This includes water boilers, kitchens, backup generators and everything else used in the building. In cold climates it means using heat pumps, in industrial buildings it includes using electric fork lift trucks, in restaurants it means all electric kitchens. Each of these comes with their own contextual challenges, cooking an Asian stir fry without a flame for example and keeping electric forklifts charged and operationally efficient.

Transitioning away from fossil fuels in is clearly a challenge but one that has to be met. The European Union Energy Performance of Buildings Directive (EPRD), adopted in May 2024, requires all new residential and non-residential buildings to be zero emissions from the start of 2028 for public buildings and 2030 for all other buildings (European Commission, n.d.). Under the EU definition, a zero emissions building is one that uses no operational fossil fuels, uses renewable energy and has "very high energy performance".

Net zero buildings may be a challenge but electrification is a significant step and not one that is beyond the reach of technology. The Hong Kong Green Building Council (HKGBC) has a rating for net zero-carbon-ready buildings that are "highly energy efficient and resilient buildings that either use renewable energy directly, or rely on a source of energy supply that can be fully decarbonised, such as electricity or district energy". For net zero buildings the HKGBC definition is "annual on-site renewable energy generation shall be equal to or more than the annual energy consumption of the building" (HKGBC, 2023).

The American Council for an Energy-Efficient Economy (ACEE) is a research organisation that helps develop policies to save energy and combat climate change. The ACEE report "Building Electrification: Programs and Best Practices" notes that while building electrification increases grid load, the demand side can be managed by smart technology, improving efficiency through passive design improvements and upgrades, battery storage and on site renewable generation. Together with grid decarbonisation, electrification will in almost all cases reduce CO2 emissions (Cohn & Esram, 2022). One question remains, who pays to upgrade the grid to meet the increased electricity demand?

Harvard University has set a goal to be fossil fuel free by 2050. Known as Goal Zero it includes all buildings and vehicles and makes the distinction between zero emissions and fossil fuel free. The target is based on the latter in recognition of the whole range of harm caused by fossil fuel production and use. This extends to eliminating the use of plastics and oil based chemicals. Building solutions in consideration include:

1. Ground-source heat pumps (GSHP)
2. Heat recovery chillers
3. Air source heat pumps
4. Steam and high temperature hot water heat pumps
5. District energy solutions
6. Building-level solutions
7. Energy Storage (Harvard Office for Sustainability, n.d.)

Among multinational companies with advanced building electrification targets, Google has stated their intentions in their 2023 Environmental Report, quoted as below:

"Electrification—which helps enable transitioning away from fossil fuels—is an essential first step toward decarbonizing our operations, followed by 24/7 CFE (Carbon Free Energy) procurement. We're exploring opportunities to electrify our facilities, from retrofitting existing building systems, to including electrification standards for new development, to working with landlords on electrification-friendly leases for buildings we don't own". They have also made use of energy audits which in Asia Pacific helped identify over 200 actions with an estimated 5600 MWh of savings (Google, 2023).

Among key challenges to achieving full building electrification in Asia are replacing backup generators and gas cooking appliances. Sustainable generator alternatives include battery and biodiesel. According to National Grid, and electricity and gas distribution company listed on the London and New York Stock Exchanges, battery energy storage systems (BESS) are an essential part of the energy transition. Lithium-ion batteries are the dominant technology which can be combined with a smart management system to optimise storage and release to help buildings manage energy peaks to reduce costs as well as serve as emergency backup in case of power cuts (National Grid, 2023). It is worth also noting the Jevons Paradox, the economic phenomenon where increased efficiency in the use of a resource result in an increase in total consumption. In this context, switching to a green energy "carbon efficient" tariff may result in slower implementation of energy saving measures and retention of more wasteful practices.

Viridi is a US-based company "pioneering fail-safe distributed energy storage". Among other products, they manufacture Lithium-ion batteries for use in commercial buildings (Viridi, n.d.). Gold Peak Technology, the world's 3rd largest battery producer, is a Hong Kong Stock Exchange listed company with a subsidiary GP Energy Tech developing nickel based battery technology for use in buildings, including data centres. Compared to lead acid batteries which have low energy density and are high maintenance or lithium-ion batteries which can overheat, nickel-zinc batteries offer high energy density and are more environmentally friendly (Gold Peak, 2023).

For all electric kitchens the challenge is partly cultural and partly technological. The traditional Asian, stove top cooking like stir frying, requires a flame. It is a challenge to imagine an all electric kitchen where stir frying is possible. The same might apply to a pizza oven where a gas or wood oven produces a drier heat than an electric equivalent. In other cooking techniques, such as steaming, it is easier to adapt. Other than that, challenges may include increasing electricity supply and retraining kitchen staff (US Department of Energy, 2023). In June 2024, both Grosvenor Estates and Lendlease committed to phasing out gas from new kitchens by 2030 (Haynes, 2024).

Refrigerants and Cooling

According to UNEP and the IEA, up to 8 years of total GHG emissions could be avoided over the next 40 years by improving standards for air conditioning (Ambrose, 2020). Gases used in refrigeration and air conditioning (RAC) equipment are typically chlorofluorocarbons (CFCs), Hydroflurocarbons (HFCs) and hydrochlorofluorocarbons (HCFCs). There is a precedent for success from The Montreal Protocol, universally adopted in September 1987 which regulates the production and consumption of man-made ozone depleting substances (ODS). These include CFCs and HCFCs. To help developing countries with compliance a fund was set up in 1991 (UNEP, n.d.). The 2007 Montreal Amendment and the 2016 Kigali Amendment sped up the phase out of HCFCs and added HFCs to the phase out list. The combined impact is an estimated prevention of 135 gigatonnes of CO2 covered by the Montreal Protocol and 105 billion tonnes under the Kigali Amendment (UNEP, n.d.).

HFCs generally have GWP of 12 to 14,800, according the UNEP OzonAction Fact sheet. R-410A, for example, developed by Honeywell in 1991 as a replacement for ozone depleting R22 (banned for use in the USA since January 1st 2020), has a GWP of 2088 and is itself now being phased out to be replaced with R32 which has a GWP of 675 but is mildly flammable. The cost implication for building owners is that to use R32 instead of R-410A the equipment itself needs to be replaced or extensively retrofitted due to the flammable nature of the gas (Raine, 2024). From January 1st 2025, refrigerants with a GWP of more than 750 are banned for use in the UK and EU (Edwards, 2023).

The scale of the problem is vast. Around 2 billion air conditioning units are in use around the world, 70% of which are for residential use (IEA, n.d.). A 2022 report published by the Institute for Governance and Sustainable Development (IGSD), Environmental Investigation Agency (EIA) and National Resources Defense Council (NRDC) reveals the scale of the problem of refrigerants. Globally, around 24 billion tonnes of CO2 equivalent are already installed in existing HFC and HCFC using equipment which compares to the approximately 35 billion tonnes of CO2 currently emitted a year (Ritchie & Roser, 2020). In the retail sector, supermarket cold supply chains account for 70% of supermarket scope 1 and 2 emissions according to data from the Environmental Investigation Agency, (Jones, 2025). These refrigerants may leak into the atmosphere in ageing or poorly installed and maintained equipment and at end of life if no recovery process is used.

The Chartered Institute of Building Services Engineers (CIBSE) estimates an annual leakage rate of 6%. Using whole life carbon assessment suggests that refrigerants can contribute up to 13% of total carbon emissions for a building using R410-A (Leiper, 2020).

The solutions to the refrigerant problems include life cycle refrigerant management (LRM):

1. Use low-GWP refrigerants.
2. Minimise use of refrigerants though passive building design and reducing length of refrigerant pipes.
3. Record top ups to monitor leaks.
4. Install gas detection equipment.
5. Implement targets for low leak rates.
6. Use best practices to install and maintain all equipment.
7. Report and enforce existing laws with modern technology.
8. Develop specialist technicians to implement best in class approaches.
9. Recover and safely dispose of 100% of refrigerant at end of life (Leiper, 2020), (Theodoridi et al., n.d.).

Returning the 2023 Google Environmental Report, refrigerants were not fully included in their calculations before 2022. In the appendices, page 87: "Beginning in 2022, in an effort to continuously implement best practice methodologies, we included fugitive emissions from refrigerant leakage in our operational boundary. These emissions represented approximately 17% of our 2022 Scope 1 GHG emissions".

Now that they have the data, Google are taking steps to address this issue. Page 37: "We're taking steps to more accurately measure refrigerant leak rates and working to develop new technology solutions that can prevent emissions from these leaks. For example, at our data centers, we're developing and deploying cooling solutions that include natural, low-GWP refrigerants".

At a national level, measures taken in Singapore by the National Environment Agency (NEA) include:

1. Use of refrigerant with GWP of less than 750 only.
2. Mandatory use of certified chiller technicians.
3. Registration of companies involved in chiller servicing works with NEA.
4. Recovery of all refrigerants with GWP of over 15.
5. Send all recovered refrigerants to NEA approved treatment facility.
6. Maintain and submit record of works.
7. Introduction of Refrigerant Handling for Chillers course to certify technicians (Singapore Government, 2025).

The Carbon Containment Lab is a not for profit organisation at Yale University with a mission to "develop and support emergent climate solutions by applying scientific, entrepreneurial, and investment expertise" (Carbon Containment Lab, 2025). In a post dated December 2022, the following points are made:

1. The number of air conditioners in use is estimated to triple by 2030.
2. R-410a is the most common gas used in air conditioning, it has a GWP of 2088 while the newer R-32 has a GWP of 675.
3. The installed refrigerant bank (refrigerant in use today) is 24 billion MTCO2e.
4. Air conditioning units leak gas during their operation and gasses are generally released at end of life.
5. An estimated 90% of refrigerants are released at end of life in the US and nearly 100% in developing countries. The release of refrigerant in Indonesia, for example, up to 2050 would release the same emissions as electricity consumption in Jakarta over the same period.
6. Refrigerant recovery and destruction is a significant opportunity for permanent mitigation which can be funded by generating carbon credits from their destruction (Chao & Mayhew, 2022).

An estimated 3 times the total 2019 global energy-related CO2 emissions could be avoided by implementing LRM in the United States alone. This is in addition to the benefits of the phasing down of high GWP refrigerants under the Montreal Protocol and Kigali Amendment (Theodoridi et al., n.d.).

Cooling Case Study

Danfoss is a Danish engineering company, founded in 1933, that has emerged as a "leading supplier of innovative and energy efficient solutions that increase machine productivity, reduce emissions, lower energy consumption and enable electrification", according to their website. Their case study "monumental energy savings in cooling systems" presents the case of a 20-storey mixed use building in Kuala Lumpur with different cooling needs at different times of the day and night.

Installation of Pressure Independent Balancing and Control Valves helped simplify the balancing procedures and reduce manpower to make adjustments. Improved technology to manage the cooling system efficiency based on actual demand together with upgrading the constant air volume system to a variable air volume system allows adjustment of air flow at different times. Energy savings of 67% with a pay back of two years are claimed as well as improved lifetime of the system (Danfoss, 2014).

Building Envelopes

Building envelopes are the physical barrier between the outdoors and the indoors, including façade, exterior walls, windows, roof and foundations. At the time of construction, the building envelope is usually expected to last as long as the building so selection of quality materials is important.

In deep building retrofits, however, the building façade may be removed and replaced while the main load supporting structure is left in place, modified or extended. The performance of the envelope is an important factor in reducing energy requirements for cooling and heating in hot and cold climates and is measured by heat loss coefficient, that is the surface area of the envelope multiplied by the thermal transmittance (Mahmoodzadeh et al., 2021). This can then be represented as a quantum of energy and a financial cost attributed to it.

Between 2010 and 2023, demand for cooling increased by two thirds and heating by 4% in advanced economies. To meet Net Zero Emissions 2050 targets, cooling and heating energy intensity need to fall by 25 and 35% respectively by 2030 (IEA, 2023). Improving building envelope performance is a crucial part of meeting NZE targets and examples are provided below to demonstrate the variety of ways these can be achieved.

Codification of building envelope performance standards should be part of this process, reducing air infiltration through the envelope is a key to reducing cooling/heating-related energy requirements. According to the IEA, in 2022 over 110 countries did not have requirements for building energy standards at all, this is equivalent to more than 2.4 billion square metres of floor space. According to the Net Zero Emissions by 2050 targets standards for zero-carbon-ready building energy codes should be set by 2030 for residential and non-residential buildings and for retrofitting of 20% of existing building floor area to this level by 2030 (IEA, 2023).

While building envelopes might sound a relatively straightforward part of a building compared to, say, refrigerants and building electrification, the University College London Building Envelope Research Network (UCL BERN) suggests otherwise. BERN incorporates research from Schools of Architecture, Engineering, Planning, Energy, Environmental Design, Sustainable Construction and Sustainable Heritage (UCL, 2025). The inter-faculty collaboration of UCL BERN offers to help the industry meet the following:

1. Bring low carbon solutions to market.
2. Identify optimised solutions that consider occupier comfort, carbon footprint, adaptability and health and safety.
3. Help occupiers achieve emission reductions.
4. Develop integrated policies (UCL, 2022).

In the United States, the US Department of Energy (DoE) initiative Better Buildings Alliance has launched the Building Envelop Campaign (BEC), a collaboration between the DoE, the American Institute of Architects (AIA), the International Facility Management Association (IFMA) and the International Institute of Building Enclosure Consultants (US Department of Energy, n.d.).

One of the BEC recognised projects is Pacific Tower, a multi-tenanted commercial building in Vancouver, Washington. Completed in 1977, the building cooling and lighting systems had been upgraded to improve energy efficiency before attention turned the windows. When full window replacement proved prohibitively expensive, secondary windows were installed at 89% of the cost and energy savings of 13.4% were achieved.

Another case cited is a Vemeer Global Parts Distribution Center, a warehouse in Iowa. Insulation improvements to walls and roof, installation of solar power generation capacity on the roof, installing thermally efficient windows and improving and air tightness resulted in building envelop performance improvement of 35%.

IXO's Lizette McNeill says the key façade improvement measures should include improving daylighting and shading, and regular replacement of gaskets to maintain integrity of building airtightness. "If these measures are adopted on a typical 30–40 year old building in Hong Kong, energy savings of 30% or more are achievable. The key to maximising cost savings is extending the life of the HVAC by upgrading the existing façade. Deep retrofitting, involving replacement of the façade will enable an efficient new HVAC compressor to deliver 50 to 80% energy savings".

Additional energy savings from façade performance improvement include spandrel insulation (spandrels and spandrel beams are panels or beams installed where the floor slab meets the façade). Without proper insulation they can result in thermal bridging of heat or cold from the exterior to the interior via the floor / ceiling slab (Walshaw, n.d.). A deeper façade retrofit option for an older building of say 25–40 years old would allow for removal of the façade, re-processing the glass panels for improved wind and solar performance, reinstallation of the glass curtain wall, introduction of temperate buffer zones to reduce the air conditioned floor area, sunshades, solar tubes to reduce lighting requirements and significant increase to solar power generation. Where permitted a wooden extension on the top the building could help offset the costs.

Façade Design Case Study

A case from Hangzhou, China undertaken by James Brearley, Principal, Brearley Architects and Urbanists (BAU) and his team is the Archives Building in Tonglu, Hangzhou completed in 2019. The building includes office, archive and laboratory uses. The façade design demonstrates both a self-shading elevation by the use of horizontal and vertical screens as well as thermal bridging insulation. The passive solar shading enables low-tint glazing, high light emittance and reduced need for lighting. The narrow building floor plate is designed to allow for cross ventilation to further reduce air conditioning loads (Figs. 7.1, 7.2, 7.3, 7.4).

Adaptive façades are environmentally friendly façade systems that work to save energy and at the same time promote user comfort by adapting to outdoor weather conditions. They can do this by integrating the façade design with the HVAC system. According to Turkish smart facade designer Toprak, this can include:

1. Integration with heat pumps.
2. Air circulation to cool and improve photovoltaic panel efficiency in summer.
3. Evaporative cooling of the façade.
4. Solar façade to maximise heat gain during winter (Toprak, n.d.).

A selection of innovative and adaptive façade concepts include:

Fig. 7.1 Archives Building, Tonglu, Hangzhou (image courtesy of BAU). Vertical screens are made of perforated aluminium to provide shade from the sun and maintain unimpeded views from within the building

Fig. 7.2 Corner detail of Archives Building, Tonglu, Hangzhou (image courtesy of BAU). Slab edges extend beyond the window line to provide shade and support for the vertical screens which will be exposed to high winds during typhoons

1. Sponge 3D that incorporates air and liquid cavities that can control heat conductivity providing thermal cavities and heat storage on different sides of the building according to external conditions and internal comfort needs.
2. GEOtube saltwater façade that evaporates to leave a solid salt white surface that continuously develops and transforms over time.

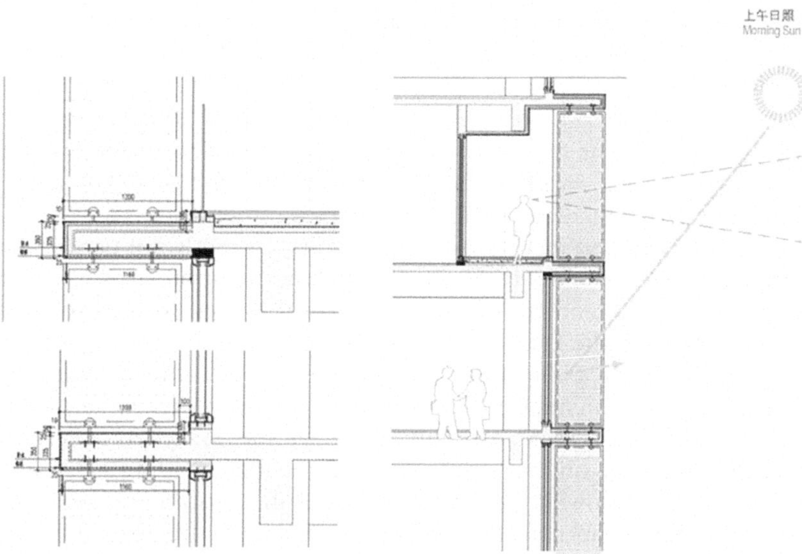

Fig. 7.3 Design detail of thermal bridging and vertical sun shade design (image courtesy of BAU). The concrete eaves are wrapped in high density insulation and weatherproofed.

3. Lattice responsive shading façade that changes on different days of the year to provide shading outside the main glass façade as installed on Al Bahar Towers in Abu Dhabi. This reduces solar gain and glare.
4. Double Face façade systems that provide a translucent layer, outside the glass, that adapts according to thermal needs to collect solar radiation in winter and deflect it in the summer.
5. Adaptive glazing that changes thermal and optical performance depending on the weather to control solar gain and daylight.
6. Algae façade that generates and stores heat and biomass.
7. Prosolve370e air cleaning façade. Titanium dioxide in the materials helps reduce air borne pollutants (Souza, 2019).
8. Vertical greenery systems (VGS) where vegetation and the soil provide shade and evapotranspiration cooling by over 15% for green façades and 8.5% for living walls (Evola & Lucchi, 2024).

9. Transparent solar panels allow daylight in while generating electricity. This is still an emerging technology but the benefits are clear.

Property management teams would at this point remind architects and façade designers that façade design needs to allow for easy maintenance and cleaning, costs increase with more complex installations.

Lizette McNeill also mentioned façade audits, currently not part of the energy audit requirements in Hong Kong, "the question is, why don't more people understand the opportunity? Perhaps they are just so used to looking through the façade they don't actually look at them". Regular inspections of the façade can not only identify structural issues but may also help identify potential energy saving opportunities. A façade audit includes "a thorough visual inspection of the façade, as well as collection and analysis of data on the materials used and the building's history" (Kapcher, n.d.). Knowing the materials used in construction and any subsequent maintenance work can help guide the audit plan to target key areas.

The Hong Kong Institution of Engineers reported on the use of drones and to check façades saving 67% of time taken and 52% of the cost while providing a high degree of accuracy for visual defects and thermal anomalies (De Filippo et al., 2020). Thermal imaging can help identify areas of energy loss through thermal bridging as well as water leaks and debonding of the structure (Barahona et al., 2021). With this information, renovation targets can be identified and prioritised.

Interior Fit Outs

Office and retail units are typically leased on a bare shell basis, with the space returned to the landlord at end of lease in the same condition. The tenant then arranges their own renovation, or fit out, to be undertaken by a project manager, architect and contractor in accordance with relevant standards and requirements set by the landlord. It can be a particularly resource intense process when carpets, furniture, IT equipment and so on are all included. It is not unknown for a fit out to be completed, tenant never move in and the fit out to be ripped out, unused, at the end of the lease.

According to RESET, a set of standards and assessment tools for healthier and more sustainable built environments, embodied carbon from office and retail renovations may exceed structural embodied carbon over the life of a building (RESET, 2022). Raefer Wallis, Founder of GIGAbase, RESET and A00 Architecture, started out with residential projects where insulation, thermal comfort and sustainability come together naturally. His transition to office interiors followed where IAQ, health and material choices also come together. Working in Shanghai, his first request for a sustainable office interior, supported by a corresponding willingness to pay, didn't come until 2019.

Stephen White recalls that his multinational clients, such as Disney English, specified minimum environmental standards for the internal finishes used in their Disney English Learning Centres. "Parent awareness made it a big marketing point as parents with only one child were very focused on health and were prepared to pay for better air quality environments". Green materials in offices happened at the same time, "US based National Instruments set the standard with recycled materials and filtration of air in their 10,000 sq.m head office, staff were able to monitor indoor air quality on their floor and compare it with external air. Staff awareness of IAQ developed generally, particularly among pregnant staff, employers were sometimes pressured to allow 'breathing periods' for new construction becoming common before move in".

Renovation materials used in interior fit outs in a typical multi-tenant office building can be replaced every 3 to 5 years under a typical lease term in Asia. On a single case they may be less significant than the building structure related materials and therefore easily overlooked both by tenants and landlord. Over a 70 year building lifespan, a 3 to 5 year tenant cycle means 14 to 20 or more refits. According to RESET Standard, a positive impact scoring methodology for building materials and projects, a typical office fit out uses 190 kg CO2 per sq.m (excluding MEP) compared to 840 kg CO2 per sq.m for a shell and core building structure (Wallis, 2022). This means the cumulative fit out emissions will be 2660 to 3800 kg CO2 per sq.m over the 70 year life cycle. Even if the same tenant remains in the building for a longer lease of, say 15 or 25 years, the fit out may undergo renewal and upgrading every 8 to 10 years.

With the RESET Standard and others like LEED Interiors, there is a means to measure and manage fit out related emissions. Many tenants, including some with Net Zero targets, simply don't include this within their measurement scope assuming that a green-certified fit out is enough. Setting procurement standards to source low carbon materials can help the emissions management process and contribute to adjustment of manufacturers own emissions as they sense the market demand for lower carbon materials. Apart from low carbon materials, other efforts to reduce fit out emissions may include circularity such as designing for recyclability, using recycled materials, diverting waste from landfill, donations of office furniture or old IT equipment.

This is happening in real life. Page 40 of the Google, 2023 Environmental Report states: "We engage with our suppliers—including hardware manufacturing and indirect services suppliers—to help reduce their energy consumption and GHG emissions, as stated in our Supplier Code of Conduct, which all suppliers are required to sign. We assess suppliers' practices to report, manage, and reduce their emissions and incorporate this into our supplier scorecard".

Material manufacturers are taking action. One that has changed its whole manufacturing process is Interface, "a global leader in modular flooring". Since 1996, they have reduced market based GHG emissions by 96% and achieved an 81% reduction in waste sent to landfill at carpet manufacturing sites. 51% of materials used in their carpets are bio-based or recyclable and they use 79% renewable energy in manufacturing (Anderson, n.d.). According to Stephen White "better products are increasing their market share as a result".

By embedding renovation related data collection into the process, appropriate measurement and management at the planning stage it is possible to significantly reduce fit out related carbon emissions. Other initiatives include reusing furniture such as Sustainable Office Solutions (SOS) in Hong Kong who work with property developers, landlords, tenants and property managers to facilitate the responsible reuse of office furniture. To date they have diverted more than 6.5 million kg of waste from landfill (SOS, n.d.).

While renovation carbon itself is often overlooked, even more may be missed by ignoring the mechanical, electrical and plumbing elements. According to RESET, these can add 30% to total embodied carbon for office interiors and are almost always excluded from calculations from other green certifications.

For tenanted buildings, landlords may offer guidance to tenants in the form of a green fit out/renovation guide. It is common for planned fit outs to require landlord approval and approval can depend on meeting these guidelines.

A typical sustainable fit out guide may include the following:

1. Green certification of fit out.
2. Sustainable design features, including:
    a. Maximise natural lighting and ventilation.
    b. 100% LED lighting.
    c. Motion sensors and lighting controls.
3. Certified sustainable contractor.
4. Energy efficiency strategy with active and passive measures.
5. Water conservation with low flow taps.
6. Circularity principles including:
    a. Contractor waste management plan to divert waste from landfill.
    b. Materials using recycled content.
    c. Recyclable materials.
7. Use only low volatile organic compound materials for floor coverings, paints and varnish, sealants, composite wood materials, furniture, fabrics and upholstery.
8. Optimisation of HVAC system.
9. Green label IT equipment.
10. Regular energy audits.

The Crown Estate Sustainability Fit Out Guide offers guidance for all fit out work that includes water efficient sanitary equipment using dual or low flush WC's, waterless urinals and low flow taps. For waste, the guide includes diverting waste from landfill, a fit out waste plan that includes options for reuse and recycling (The Crown Estate, n.d.).

"In many cases, such fit out guides are only the start and could, over time, fundamentally change the market", according to Wallis. "Tenants simply cannot achieve their targets without stronger support from landlords, the leading landlords have already recognised that providing this support creates significant competitive value in a challenging leasing market", Wallis added. Specifying a carbon footprint, detailed carbon accounting boundaries, requirements for disassembly and a circularity plan, etc., may become a standard part of a green lease agreement. (Fig. 7.4).

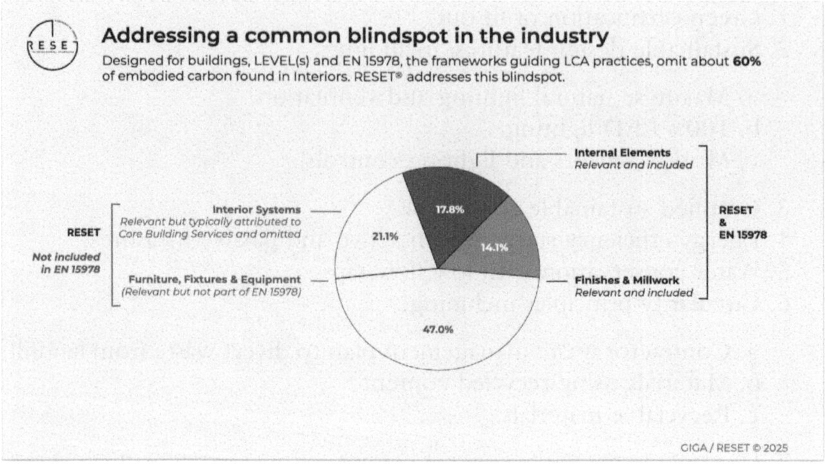

Fig. 7.4 Sources of Embodied Carbon in Fit Outs (RESET, 2025)

Water

A recurring theme in throughout this book that hasn't yet been addressed in detail is water. A starting point is perhaps the relationship of water, building value and liquidity. According to property agents Knight Frank in July 2023, "homes by the water were worth 48% more on average than the equivalent non-water front properties in Q2 2023" (Bailey, 2023). Not only that, the same report states that 54% of buyers of waterfront property in the UK paid cash compared to around 29% overall. These properties are a more liquid asset than others.

This should not be a surprise. Students of history or geography will know the relevance of the Tigris and Euphrates rivers in the development of Mesopotamia as the cradle of civilisation, the role of the Nile in Ancient Egypt, the Tiber in Rome, the Yangtze in China, the Ganges in India, the Mekong that supports 20,000 plant species, 850 fish species and around 65 million people in the Lower Mekong Basin alone (Mekong River Commission, 2024). The mightiest of them all is the Amazon River Basin that covers more than 35% of the entire South American continent.

Water exists in our consciousness as desirable and valuable but is rarely considered in the context of the built environment. Yet the built environment depends on it and then discards it with barely a second thought. Water is used in all stages of construction and building operations, including wastewater and sewage. 15% of freshwater is used in building operations and construction with 30–40% of that used on toilet flushing alone (WSP, 2024).

The LVMH Life in Architecture guide includes a water stewardship section. Key highlights are:

1. Any LVMH project with outdoor areas must include permeability meaning a green roof, a rain garden, permeable paving, rainwater collection to meet a permeability standard of 60% where there is no local require standard.
2. Rainwater recovery aims to reduce the use of tap water for toilets and landscape irrigation. A particular encouragement to work with the landlord to do this.
3. Water measurement applies to any project with a water supply broken down into specific uses, toilets, gardens, maintenance and divided into drinking and non-drinking water. A saving of more than 20% may earn the top 3 stars.
4. Fixed target criteria for water efficient taps and sanitary fittings based on litres per minute. Over 90% compliance earns a top 3 stars (LVMH, 2025).

A Savills blog article offers guidance for a "portfolio-wide water efficiency strategy". This starts with collecting consumption data based on water saving measures already in place, establishing a list of actions to improve water conservation and additional data collection as required and measuring ongoing consumption patterns. The key elements of the strategy include:

1. Site-specific actions based on consumption patterns, cost, life cycle stage of the building, building refurbishment plans, tenant and stakeholder expectations.
2. Impact based prioritisation. This includes considerations such as landlord control, tenant collaboration, high consumption facilities and potential to reduce consumption.
3. Alignment with an industry framework for water efficiency. This includes green certifications that provides water conservation actions to be measured (Chantry, 2023).

According to Water Intelligence (WINT) a business that helps manage water resources in buildings to reduce both operating costs and carbon footprint, each cubic metre of water consumed produces over 10.5 kg of carbon emissions (Dycian, 2022). In the US, buildings waste around 25% of water through cooling tower problems, leaking toilets and other leaks. Escape of water can result in structural damage, repairs, loss of other services, e.g. electricity and mould leading to health risks. According to the Association of British Insurers, domestic property alone results in GBP1.8 million per day in UK insurance claims (ABI, n.d.). Effective water preventative maintenance, consumption monitoring, leak detection and so on are all part of reducing costs and carbon emissions. If you are building new or retrofitting existing now to last the next 50 to 100 years, you should be doing everything you can to minimise water consumption and maximise water capture and recycling.

Water stress is becoming more severe as resources are depleted and climates change. It is not difficult to manage by monitoring consumption through use of metres and improving management of high consumption activities. Benchmarking and target setting can be used once the consumption patterns are understood. Leak monitoring can also reduce water loss (The Crown Estate, n.d.). In water stressed cities, water conservation, leak reduction, rainwater harvesting and grey water recycling are all about resilience and cost savings.

Beyond buildings, in the broader context of the built environment, China launched the Sponge City initiative in 2015 as part of an initiative to combat more severe and frequent flooding in our growing cities. The aim set was for 80% of urban areas in China to use 70% of their rainwater (WEF, 2017). Achieving this requires nature based solutions, including trees, lakes, parks, rain gardens, permeable hard surfaces, roof top gardens and so on to capture and store rainwater in the natural environment within cities until it can be drawn on for use. This is a transformation from the rapid urban drainage model which involves discard of water into the nearest river, lake or sea and results in a lost resource (WEF, 2022).

Joe MacDonald, author of the Net Zero Carbon Strategist update, confirms the role of sponge cities as a means of rethinking urban water management. He describes the benefits of sponge cities as including stormwater capture, flood control, urban cooling, increased property values, new jobs and carbon sequestration (MacDonald, 2025).

In Bangkok, Benjakitti Park is a spectacular 720,000 sq.m. wetland Forest Park built on the site of a former tobacco factory in the city centre.

The "sponge" lies at just 0.5 metres above sea level and has a capacity to absorb 128,000 cubic metres of water in the rainy season and uses bioremediation to deliver 1600 cubic metres of clean water a day (Abdel, 2023). According to the World Bank, 40% of Bangkok could be flooded by 2030 and it tops the Intergovernmental Panel on Climate Change (IPCC) list of the cities at most risk of sea level rise flooding impact (Igini, 2022) (Fig. 7.5).

Even where water is abundant, there can be issues with water quality. Flint in Michigan, USA is one of the best known recent cases. Flint River had long been used as a dumping ground for industrial waste from car assembly, meat packing and paper mills as well as raw sewage. Not surprisingly, when the city switched from the Detroit water supply to Flint River water to save costs, poor treatment and inadequate testing led to health conditions from the poor water quality (Denchark, 2018). In the EU, new regulations limiting "forever chemicals", polyfluoroalkyl (Pfa) substances in drinking water from January 2026 are expected to result in tap water bans across Europe (Weston, 2025). The UK has seen record amounts of raw sewage released into England's rivers and coastal waters resulting in only 14% of rivers and lakes scoring a good ecological status (BBC, 2025).

Fig. 7.5 Benjakitti Park, Bangkok (authors own photograph)

Energy Audits

Energy consumption is the most significant source of emissions in building operations. Optimising energy use throughout a building requires quality data about exactly where the energy is being used. The starting point is an energy audit, with the data in hand, analysis involving the property management and engineering teams can identify the "hot spots" using any available historic data. Interviews with tenants can try to identify issues in their tenanted areas and if available obtain energy audits for tenants space as well to analyse together with the common areas. Energy reduction options based on the energy consumption data can be detailed and an action plan developed based on the potential energy saving and cost analysis.

A typical energy audit seeks to identify all energy saving opportunities in the building and should include:

1. List and details of all the current energy consuming equipment and controls in the landlord controlled areas.
2. A building specific list of high, low and no cost energy saving improvements that can be made.
3. Potential energy savings (kWh), carbon savings and a return on investment, where investment is required, should be listed.
4. Analysis of site-specific energy use of current and historic half hour metre readings.
5. An itemised list of total savings achievable if measures are implemented.

This can be prepared at the same time as an assessment for potential renewable energy generation options. In addition, energy audits of tenanted areas conducted at the same time, and on a similar basis to the above, can identify additional savings for the tenant. Provision for landlord and tenant energy audits, metering and data sharing can be included in any green lease agreement (see section below).

An alternative approach is for the property management team and building engineers to work directly with tenants to implement energy saving measures in tenants occupied spaces. This should be considered similar to a car service that is typically done annually to ensure optimum performance. An inspection by qualified teams to focus on the condition of key energy using equipment including the air conditioning system, server rooms and so on. The inspection includes checking the metre and exactly what is included in the measurement.

Such an inspection might reveal, for example, that fan coil units are set to turn on at 05:00. This can comfortably be pushed back when office staff generally turn up from 08:30. The installed ducting might need attention to improve delivery of cold air by reducing any turbulence. In some cases drain pumps might have been installed to manage water discharge rather than actually fixing the problem. Equipment in server rooms, such as battery backup systems, might be in need of replacement or reconfiguration. In one 500 sq.m office, these things may appear insignificant, but when applied across a whole building, a whole office district, a whole city or nationwide, the impact can be significant.

The Rocky Mountain Institute (RMI) was founded in 1982 with a data led energy efficiency mission to transform the way America generates and uses electricity. Their 2018 report "Best Practices for Leasing Net Zero Energy Buildings" recommends the following actionable steps:

1. Energy budget for each tenant and base building with sub-metering to monitor performance. Commitment for those exceeding the budget to make improvements and buy Renewable Energy Certificates (RECs).
2. Base building energy performance optimisation each year to ensure all systems are operating at maximum efficiency.
3. RECs used to make up for any shortfall in renewable energy supplies.
4. Transparent reporting of building and tenant energy consumption including any on site renewable energy generation. This will enable action to be taken to get back on track.
5. Cost recovery wording to cover additional costs for investment in energy improvements and similar (Carmichael & Petersen, 2018).

The HSBC published report "Decarbonising the Built Environment, Perspectives from Hong Kong and Mainland China" recommended publishing energy efficiency data, specifically EUI based on 5 yearly energy audits. It also suggested improving access to green energy tariffs and updating building codes to permit solar panel installation. Finally encouraging a retrofitting market linked to green building standards and updates to the Building Codes and Planning Guidelines with targeted incentives and support (HSBC, 2025).

Green Lease Agreements

Green lease agreements are a means of formalising sustainability collaboration between landlords and tenants. According to Savills "a green lease is one where landlord and tenant agree sustainability-linked targets and measurable key performance indicators as part of the lease agreement. The lease might put requirements on either the landlord or tenant, or may lock in collaboration between the two" (Savills, 2022). In many cases they are not part of the lease contract at all and may be a less formal voluntary partnership.

Where they are legally binding and included in the lease agreement, the requirements may relate simply to the tenant meeting the requirements of any building green certification and compliance with relevant environmental law. In cases where they are voluntary and contained in a side agreement, they may be more specific and on a non-binding basis. This might include specific requirements on electrification, refrigerants, bio diversity on the landlords commitment side and set sustainable fit out standards, such as 100% LED use, energy efficiency standards, participation in building recycling schemes and similar on the tenant commitment side.

A key point being establishment of an ongoing dialogue between landlord and tenants. This can be achieved through a building ESG committee with terms of reference setting out the purpose of the committee, committee chair and membership, frequency of meetings, sharing of meeting minutes and documentation and so on. In Hong Kong, Swire Properties, Hong Kong Land, Hang Lung Properties and Nanfung are some of the leading commercial landlords with sustainability partnerships, each with slightly different arrangements. In other cases, it may be the tenant that has a policy to have green lease agreements in place across their global portfolio.

One of the difficulties might be finding the common ground to prioritise. While tenants may be more focused on wellness of their colleagues, such as indoor air quality, natural light and so on, landlords may want to see improvements to energy efficiency involving changes to the tenants leased space or more sustainable fit outs. This is where a building ESG committee, established in the green lease agreement, can be helpful. One role is to find compromise through communication and collaboration.

Common clauses in a green lease agreement and other documentation cover the following:

1. Building electrification to remove all fossil fuels from the building.
2. Waste management.
    1. Recyclables
    2. Food waste
    3. E-waste
    4. Waste data collection
3. Low-GWP refrigerants.
4. Building optimisation.
5. EV charging stations.
6. Maximise on site energy generation.
7. After hours lighting policy.
8. Green-certified cleaning and pest control contractors.
9. Bio diversity:
    1. 100% native plant species.
    2. Green rooves.
    3. Bird strike mitigation.
    4. Wildlife friendly lighting.
10. Sustainable fit out guide.
11. Healthy food options available in the building.
12. Community engagement and social sustainability.
13. Single use plastic free building guidance.
14. Water conservation though air conditioning condensate collection, grey water recycling and rain water capture.
15. Disaster resilience and business continuity plan.

Each of these can be prepared by each individual tenant and the landlord but by both sides working together in a way they may not be used to, they become more effective and efficient.

Sustainable Building Operations Case Study

NWP Property is a joint venture between PT City Retail Developments and Warburg Pincus established in 2015 and owns a network of around 50 community shopping centres and logistics warehouse properties. Dhawal Doshi is Head of Research and ESG at NWP Property, together with senior management implemented a mission driven and data led approach to sustainable building operations which resulted in significant energy cost savings. NWP has been recognised by the IFC for their use of ESG Certification (see chapter 9) of retail malls in Indonesia (Farrell & Hlatky, 2023).

According to Doshi "while we approached this initially as a 'nice to have' a lot of work went on behind the scenes to prove the cuts to operating costs that could be delivered. While, a lot of technical skills were required but half of the job is about knowing how to plan and deliver strategically. This required internal resources and external consultants. Many people have to be aligned, including upskilling building engineers and other professionals and a lot of this work remains invisible to external stakeholders. We started this journey back in 2017 so by 2021 when external stakeholder focus about ESG reporting increased we were able to give them all the answers and supporting evidence. Our sustainability strategy has meant that we have secured a World Bank sustainability-linked loan (the first in Indonesia) among others, increased our pool of investors by demonstrating management excellence and we avoided the risks of 'missing out'. Finally, measuring the financial returns is not straightforward. There are performance efficiency improvements in chillers that result in a more comfortable indoor temperatures as well as cost savings and each asset needs its own tailored plan. Without quantifying the indoor environment improvements, there is no possibility of an apple to apple comparison" (Table 7.1).

In addition to the above R22 refrigerants have been disposed of and grey water recycling measures have been implemented.

Sustainable Design Case Study

Sustainable design was mentioned in the last chapter in the context of reducing embodied carbon. It also helps reduce operating carbon. In Guangzhou, the 71 storey Pearl River Tower was completed in 2013.

Table 7.1 Summary of selected NWP asset sustainability performance improvements

Building	*Strategy*	*Result*	*Comment*
The Park Sawangan, Jakarta. Greenfield development	Changed design for oi-based chiller to electric. Other improvements to chiller controls, variable speed drives (VSDs), scale cleaning system, building envelope improvements and lighting changes	30% energy efficiency saving	Knowledge transfer from the design phase of this greenfield development project benefited other properties in the portfolio
Citimall Cianjur II. Existing building	Chiller retrofit	Energy savings and enhanced cooling performance	Retrofit of a 10-year old HVAC system doubles cooling efficiency
Citimall Cilegon. Older-generation acquisition	Retrofitted old split-duct AC system into a water-cooled magnetic bearing chiller	Reduction in operating costs and improvement of cooling efficiency by around 60%	Considerable upfront investment required. Detailed analysis to measure impact. Process informed approach to similar work at other assets

(Farrell & Hlatky, 2023)

It is the largest radiant-cooled building in the world. Building orientation and aerodynamic form enable it to funnel prevailing winds into each mechanical floor and generate energy. The combination of radiant panel cooling and chilled beams ceiling system means less ductwork is required than a typical office building. This enabled the architects to design five additional floors within the required height limits (SOM, n.d.).

The building design delivers a 30% energy saving compared to the Chinese energy code baseline and was among the first supertall LEED Platinum certified buildings with sustainable features including:

1. Double skin exterior façade.
2. Integrated photovoltaics.
3. Vertical axis wind turbines.
4. Daylight harvesting reflectors.

5. Daylight-responsive controls.
6. Energy efficient lighting.
7. Radiant cooling and underfloor ventilation.
8. High-efficiency chiller with heat recovery.
9. Double-deck elevators (Tomlinson et al., 2014).

This was a trailblazing project. An architect involved in the building concept mentioned the challenges of stakeholder management both in terms of delivering the project within budget, justifying the additional investment and post-completion cost recovery for occupants and the developer.

Summary

This chapter has covered the key hard to tackle sustainability performance factors in building operations. There are multiple things to consider, including refrigerants and spandrel insulation. With good planning and design they are not necessarily all that hard to tackle either from the ground up or by retrofit. instead they need to be recognised, understood through quality data and have a forward thinking and costed plan in place to implement over a period of time. The case studies described show that this is not just theory, it is management excellence.

Electrification, refrigerants, façades, water management and energy audits can all be tied up into a green lease agreement. Legally binding or not, this framework is a cost effective means to drive collaboration among landlord, tenants and property managers to improve sustainability performance.

Many of the topics discussed in this and the previous embodied carbon chapter are neatly tied up in Green Building Certification which is mentioned throughout the book but is the subject discussed in detail in Chapter 9. The next chapter looks deeper at some of the co-ordination and collaboration already taking place across the industry and the role of various professional industry associations.

References

Abdel, H. (2023, August 3). *Benjakitti Forest Park / Turenscape + Arsomsilp Community and Environmental Architect*. ArchDaily. Retrieved April

13, 2025, from https://www.archdaily.com/1004947/benjakitti-forest-park-arsomslip-community-and-environmental-architect

ABI. (n.d.). *Burst pipes and water leaks*. ABI. Retrieved July 2, 2025, from https://www.abi.org.uk/products-and-issues/choosing-the-right-insurance/home-insurance/burst-pipes-and-water-leaks/

Ambrose, J. (2020, July 17). Air conditioning curbs could save years' worth of emissions – study. *The Guardian*. https://www.theguardian.com/environment/2020/jul/17/air-conditioning-curbs-could-save-years-worth-of-emissions-study

Anderson, R. (n.d.). *Interface carpet tile & resilient flooring sustainability overview*. Interface. Retrieved April 12, 2025, from https://www.interface.com/CA/en-CA/sustainability/sustainability-overview.html

Bailey, L. (2023, July 17). *The enduring appeal of waterfront living*. Knight Frank. Retrieved April 12, 2025, from https://www.knightfrank.com/research/article/2023-07-17-the-enduring-appeal-of-waterfront-living

Barahona, B., Buck, R., Okaya, O., & Schuetz, P. (2021). *Detection of thermal anomalies on building facades using infrared thermography and supervised learning*. Retrieved April 12, 2025, from https://iopscience.iop.org/article/https://doi.org/10.1088/1742-6596/2042/1/012013/pdf

BBC. (2025, March 27). *How much sewage is spilled into rivers, lakes and the sea near you?* BBC. Retrieved July 2, 2025, from https://www.bbc.co.uk/news/explainers-62631320

Chan, C. (2024). *Strategies for driving carbon neutrality from the perspective of the HKSARG and the building industry [Seminar]*. City University.

Carbon Containment Lab. (2025). *About the lab*. Retrieved March 15, 2025, from https://carboncontainmentlab.org/who-we-are

Carmichael, C., & Petersen, A. (2018). *Best practices for leasing NZE Buildings*. Retrieved April 9, 2025, from https://rmi.org/insight/best_practices_for_leased_nze/

Chantry, W. (2023, September 11). *How to develop a portfolio-wide water efficiency strategy*. Retrieved August 28, 2025, from https://www.savills.co.uk/blog/article/351234/commercial-property/how-to-develop-a-portfolio-wide-water-efficiency-strategy.aspx

Chao, T., & Mayhew, C. (2022, December 21). *Managing refrigerants in a warmer world*. Carbon Containment Lab. Retrieved April 12, 2025, from https://carboncontainmentlab.org/updates/posts/managing-refrigerants-in-a-warmer-world

Cohn, C., & Esram, N. W. (2022, February 3). *Building electrification: Programs and best practices*. Retrieved April 5, 2025, from https://www.aceee.org/research-report/b2201

Danfoss. (2014, August 8). *Hydronic balancing for cooling in Consplant tower, Kuala Lumpur, Malaysia*. Retrieved March 1, 2025, from https://www.danfoss.com/en/service-and-support/case-stories/dhs/hydronic-balancing-for-cooling-in-consplant-tower-kuala-lumpur-malaysia/

De Filippo, M., Asadiabadi, S., Kuang, J., Mishra, D. K., & Sun, H. (2020). *AI-powered inspections of facades in reinforced concrete buildings*. HKIE. Retrieved March 18, 2025, from https://www.hkie.org.hk/hkietransactions/upload/2023-02-03/THIE-2020-0023.pdf

Denchark, M. (2018, November 8). *Flint water crisis: Everything you need to know*. NRDC. Retrieved July 2, 2025, from https://www.nrdc.org/stories/flint-water-crisis-everything-you-need-know

Dycian, Y. (2022, February). *The carbon footprint of water*. Retrieved July 2, 2025, from https://wint.ai/blog/the-carbon-footprint-of-water/

Edwards, K. (2023, May 28). *R410a refrigerant phaseout*. Retrieved March 18, 2025, from https://www.bedfordshireaircon.co.uk/r410a-refrigerant-phaseout

European Commission. (n.d.). *Nearly-zero energy and zero-emission buildings*. Retrieved April 9, 2025, from https://energy.ec.europa.eu/topics/energy-efficiency/energy-efficient-buildings/nearly-zero-energy-and-zero-emission-buildings_en

Evola, G., & Lucchi, E. (2024). Thermal performance of the building envelope: Original methods and advanced solutions. *Buildings*, *14*. https://doi.org/10.3390/buildings14082507

Farrell, M., & Hlatky, P. (2023, June 30). *Why data matters*. Retrieved August 11, 2025, from https://edgebuildings.com/why-data-matters-an-interview-with-dhawal-doshi/

Gold Peak. (2023). *Our story – Gold Peak Technology Group Limited*. Gold Peak Technology Group Limited. Retrieved April 12, 2025, from https://www.gpet.com/

Google. (2023). *Environmental report 2023*. Retrieved March 1, 2025, from https://sustainability.google/reports/google-2023-environmental-report/

Hang Lung Properties. (2025, March). *Our journey to net zero*. Retrieved April 10, 2025, from https://www.hanglung.com/getmedia/c22fa7da-a67b-49e3-a09b-f70f4b6b39c8/Our-Journey-to-Net-Zero-EN.pdf

Harvard Office for Sustainability. (n.d.). *Harvard's path to fossil fuel-free by 2050*. Retrieved April 1, 2025, from https://sustainable.harvard.edu/fossil-fuel-free-by-2050/

Haynes, L. (2024, June 4). *Major landlords serve up electric-only kitchen challenge to tenants*. Retrieved April 12, 2025, from https://www.costar.com/article/1643563670/major-landlords-serve-up-electric-only-kitchen-challenge-to-tenants

Hong Kong Green Building Council. (2023). *Build and be zero carbon ready and net zero ready*. Hong Kong Green Building Council. Retrieved April 12, 2025, from https://zcrbc.hkgbc.org.hk/

HSBC. (2025). *Decarbonising the built environment* [Perspectives from mainland China and Hong Kong].

IEA. (n.d.). *Cooling - IEA*. International Energy Agency. Retrieved April 12, 2025, from https://www.iea.org/energy-system/buildings/cooling

IEA. (2023, July 11). *Building envelopes - IEA*. International Energy Agency. Retrieved April 12, 2025, from https://www.iea.org/energy-system/buildings/building-envelopes

Igini, M. (2022, June 4). *Sea level rise projections: 10 cities at risk of flooding*. Retrieved April 11, 2025, from htthttps://earth.org/sea-level-rise-projections/

Jones, S. (2025, July 1). *Majority of supermarket carbon emissions come from fridges*. The Grocer. Retrieved August 23, 2025, from https://www.thegrocer.co.uk/news/refrigeration-accounts-up-to-70-of-supermarket-carbon-emissions/706339.article

Kapcher. (n.d.). *What is a Facade audit?* Retrieved March 21, 2025, from https://kapcher.com.au/blog/what-is-a-facade-audit

Leiper, A. (2020). *Refrigerants and their contribution to global warming*. Net zero carbon guide. Retrieved April 12, 2025, from https://www.netzerocarbonguide.co.uk/guide/designing-and-building/heating-your-building/refrigerants-and-their-contribution-to-global-warming

LVMH. (2025, May). *Life in architecture*. Retrieved August 22, 2025, from https://www.lvmh.com/commitment-in-action/for-the-environment/life-in-stores-awards

MacDonald, J. (2025, April 11). *Water as leverage: The economic and environmental impact of sponge cities*. Retrieved April 11, 2025, from https://www.linkedin.com/pulse/water-leverage-economic-environmental-impact-sponge-cities-macdonald-1tq2e/

Mahmoodzadeh, M., Gretka, V., Blue, A., Adams, D., Dallimore, B., & Mukhopadhyaya, P. (2021). Evaluating thermal performance of vertical building envelopes: Case studies in a Canadian university campus. *Journal of Building Engineering*, *40*. https://doi.org/10.1016/j.jobe.2021.102712

Mekong River Commission. (2024). *Mekong River Basin*. Retrieved March 28, 2025, from https://www.mrcmekong.org/mekong-river-basin

National Grid. (2023, May 9). *What is battery storage?* National Grid. Retrieved June 30, 2025, from https://www.nationalgrid.com/stories/energy-explained/what-is-battery-storage

Raine, F. (2024). *R410A*. National Refrigerants Ltd. Retrieved April 12, 2025, from https://nationalref.com/products/r410a/

RESET. (2022, April). *Embodied Carbon and Circularity in Office Interiors*. Retrieved May, 2025, from https://www.linkedin.com/posts/raefer-wallis_reset-embodied-https://reset.build/resources/research/embodied-carbon-and-circularity

Ritchie, H., & Roser, M. (2020). *CO_2 emissions*. Our World in Data. Retrieved April 12, 2025, from https://ourworldindata.org/co2-emissions

Savills. (2022, May 6). *Asia a slow starter in the green lease stakes*. Retrieved April 8, 2025, from https://www.savills.com/prospects/themes-asia-a-slow-starter-in-the-green-lease-stakes.html

Singapore Government. (2025, January 3). *Reducing Greenhouse Gas (GHG) emissions from the use of refrigerants in RAC sector*. NEA. Retrieved April 12, 2025, from https://www.nea.gov.sg/our-services/climate-change-energy-efficiency/climate-change/reducing-ghg-emissions-from-the-use-of-refrigerants-in-rac-sector

SOM. (n.d.). *Pearl River Tower*. Retrieved August 22, 2025, from https://www.som.com/projects/pearl-river-tower/

SOS. (n.d.). *Your SOS data*. Retrieved March 28, 2025, from https://sos-action.com/your-sos-data

Souza, E. (2019, July 24). *10 Solutions for adaptive walls for more resilient architecture*. Retrieved 21, March, from https://www.archdaily.com/920979/10-solutions-for-adaptive-walls-for-more-resilient-architecture

The Crown Estate. (n.d.). *Sustainability fit out guide*. Retrieved March 28, 2025, from https://www.thecrownestate.co.uk/media/1748/sustainability-fit-out-guide-offices.pdf

Theodoridi, C., Hillbrand, A., Starr, C., Mahapatra, A., & Taddonio, K. (n.d.). *The 90 billion ton opportunity*. Retrieved March 1, 2025, from https://www.igsd.org/wp-content/uploads/2022/10/Refrigerant-Lifecycle-FullReport-5SinglePages-PRESS.pdf

Tomlinson, R., Baker, W., Leung, L., Chien, S., & Zhu, Y. (2014). *Case Study: Pearl River Tower, Guangzhou*. Council on Tall Buildings and Urban Habitat. Retrieved August 22, 2025, from https://global.ctbuh.org/resources/papers/download/1629-case-study-pearl-river-tower-guangzhou.pdf

Toprak. (n.d.). *Smart Facade*. Retrieved March 18, 2025, from https://www.akillicephe.com/

UCL. (2022). *UCL BERN briefing*. Retrieved March 11, 2025, from https://www.ucl.ac.uk/bartlett/building-envelope-research/sites/building_envelope_research/files/ucl_bern_briefing_march_2022_6.pdf

UCL. (2025). *The Bartlett faculty of the built environment*. Retrieved March 2, 2025, from https://www.ucl.ac.uk/building-envelope-research/who-we-are

UNEP. (n.d.). *About Montreal Protocol*. UNEP. Retrieved April 12, 2025, from https://www.unep.org/ozonaction/who-we-are/about-montreal-protocol

US Department of Energy. (n.d.). *Building envelope campaign*. Retrieved March 12, 2025, from https://betterbuildingssolutioncenter.energy.gov/alliance/technology-campaigns/building-envelope-campaign

US Department of Energy. (2023, February 1). *Electrifying commercial kitchens across sectors | Better buildings initiative*. Better Buildings Solution Center. Retrieved April 12, 2025, from https://betterbuildingssolutioncenter.energy.gov/beat-blog/electrifying-commercial-kitchens-across-sector

Viridi. (n.d.). *RPS1200 large scale energy storage*. Retrieved March 15, 2025, from https://viridiparente.com/products/1-2-mwh-series/

Wallis, R. (2022, December 12). Retrieved March 25, 2025, from https://reset.build/newsroom/320

Walshaw, E. (n.d.). *Understanding thermal bridging and PSI values*. Retrieved March 18, 2025, from https://hyperion.ornl.gov/puhttps://detail-library.co.uk/understanding-thermal-bridging-and-psi-values/

WEF. (2017, November 16). *China is fighting flooding with 'sponge cities'*. Retrieved April 10, 2025, from https://www.weforum.org/stories/2017/11/china-is-building-30-sponge-cities-to-combat-climate-change/

Weston, P. (2025, July 1). *'Even if we stop drinking we will be exposed': Parts of France have banned tap water. Is it a warning for the rest of Europe?* Retrieved July 2, 2025, from https://www.theguardian.com/environment/2025/jul/01/pfas-forever-chemicals-water-contamination-saint-louis-france-aoe

WSP. (2024, August 14). *The world is facing a water crisis. Here's how the built environment can help*. Retrieved April 8, 2025, from https://www.wsp.com/en-hk/insights/the-world-is-facing-a-water-crisis-here-is-how-the-built-environment-can-help

CHAPTER 8

Industry Co-ordination and Collaboration Model

"I can do things you cannot, you can do things I cannot; together we can do great things". Mother Teresa

Who Is Driving Change and Why?

So far, we have looked at the business case for sustainability in the built environment, the role of cities, the ways in which buildings contribute to emissions and climate change, the various key material and operational inputs and their environmental impact. In this chapter it's time to start looking more deeply into the role of co-ordination and collaboration in the built environment value chain in delivering net zero, how the industry is organised, the industry associations and the drivers for change in both new construction and existing built environment. Most importantly some case studies showing the extent to which an industry governance model is starting to take shape, as it must (Fig. 8.1).

Bear in mind out the outset the sample 40-year-old 50-storey mixed use commercial building in Hong Kong. This is a non-core CBD location, as an office cluster it performs well but building owners probably lack the tenant demand and direct revenue relationship" with sustainability to undertake a deep retrofit that they might in a core-Central CBD location. In short, it represents the "existing building dilemma" of how to

S. Crispin, *Sustainability in the Built Environment*,
https://doi.org/10.1007/978-981-95-6549-8_8

Fig. 8.1 Building under construction, Pune, India (authors own photograph)

retrofit with limited resources and fewer occupier-led drivers than core CBD locations.

Given the volume of emissions related to the built environment as mentioned in previous chapters, it is easy to see why Professor Chris Goodier of Loughborough University School of Architecture, Building and Civil Engineering says "climate change and the route to net zero is the greatest challenge of our generation – with the built environment being a significant and intrinsically important part of this journey" (Goodier, n.d.).

According to Mark Reynolds, Group Chairman and Chief Executive of Mace, a construction firm, reducing emissions in construction requires a change of mindset and includes the role of legislation, creation of a carbon fluent workforce through upskilling (project managers, lawyers, engineers, marketers and others), and the whole industry rising to meet the challenge ahead (Reynolds, 2024).

Views on sustainable buildings are going mainstream among industry leaders. At JLL, a real estate agency, President and Global CEO Christian Ulbrich was quoted in May 2024 saying "I think we are still at the early stages of people realizing that there will be no future for buildings which are not green… So you are either able to turn your existing building into a green building… or the value of your building compared to neighbouring green buildings will just kind of melt away" (Wilmot, 2024).

At Cushman & Wakefield, another commercial real estate services company, achieving net zero and beyond in the built environment means "bringing together sustainable design, efficient use of resources and inclusivity". They highlight six strategies to go beyond net zero to "do good":

1. Sustainable design.
2. Net zero energy and carbon neutral.
3. Healthy design and operation.
4. Social equity, community engagement, education and awareness.
5. Data, technology, policy and regulation.
6. Regenerative design (Cushman & Wakefield, n.d.).

What Does Net Zero Mean for the Built Environment?

What exactly is net zero? Picking a definition requires an authoritative source and simplicity. The United Nations definition is as follows: "net zero means cutting carbon emissions to a small amount of residual emissions that can be absorbed and durably stored by nature and other CO2 removal measures, leaving zero in the atmosphere" (UN, n.d.).

It is useful to remind ourselves of the key difference between net zero and carbon neutrality. Net zero starts with a focus on making all feasible emissions reductions and then balancing the remainder with offsets to achieve net zero. Carbon neutrality means balancing emissions with offsets without first making all feasible emissions reductions.

For the built environment to achieve net zero-carbon, it must minimise whole life emissions, that is both embodied carbon from materials and operational carbon from day to day use, while at the same time ensuring future-proofed resilience. This means being as low carbon as possible across the life cycle which covers:

1. Material extraction
2. Production of building materials
3. Transportation
4. Construction process

5. Operational energy
6. Building maintenance and renovations
7. Interior decoration and fit out
8. Occupier behaviour
9. Demolition
10. Waste disposal (GlobalABC, 2021)

In practice, for existing building operations, this means going fossil fuel free to be fully electric and then using 100% renewable energy. Achieving net zero in the built environment globally means doing this and implementing a low carbon supply chain for the list above, across all buildings in all locations. Quite a hurdle but at the same time, mission critical.

Industry Collaboration Model

While buildings themselves are complex, the network of built environment industry-related organisations is, if anything, even more complex. Overlapping and interlinked groups, research led, policy focused, collaboration harnessing, action catalysing, theory of change driving to the extent that getting lost and befuddled is almost inevitable. The following section aims to summarise some of the leading organisations, their aims and their actions based on publicly available information.

The global body for built environment stakeholders committed to a common vision of "zero-emission, efficient and resilient buildings and construction sector" is Global Alliance for Buildings and Construction (GlobalABC), hosted by the United Nations Environment Programme. GlobalABC describes itself as "a voluntary partnership of national and local governments, inter-governmental organisations, businesses, associations, networks and think tanks committed to a common vision: A zero-emission, efficient and resilient buildings and construction sector." It was launched at the 21st Conference of Parties (COP21) in Paris, 2015 and aims to be:

1. An advocate and catalyst for global action by defining a carbon neutrality strategy for the built environment.
2. A trusted platform for tracking progress against targets including status reports, decarbonisation tracker and as co-leader for the

Human Settlements Pathway as part of the Marrakech Partnership for Global Climate Action.

3. Supportive to countries setting measures and priorities relevant to their local context.
4. Supportive to the transition to decarbonisation and increased resilience in the private sector (GlobalABC, n.d.).

Overcoming key barriers to the vision involves:

1. Awareness and education.
2. Public Policies.
3. Market Transformation.
4. Finance.
5. Building measurement, Data and Information.

The Breakthrough Agenda was launched at COP26 (Glasgow 2021) to close the "collaboration gap" among national governments, businesses and civil society to speed up actions towards the Paris Agreement decarbonisation targets. Aims include collaboration to make sustainable solutions and clean technology the most attractive, accessible and affordable options by 2030 (The Breakthrough Agenda, n.d.). Buildings Breakthrough has a target to make near-zero emission and resilient buildings the new normal by 2030. 28 countries have signed up to the Buildings Breakthrough: Armenia, Austria, Canada, China, Cote d'Ivoire, Egypt, Ethiopia, Finland, France, Germany, Ghana, Guinea-Bissau, Japan, Jordan, Kenya, Liberia, Mauritania, Mongolia, Morocco, Netherlands, Norway, Senegal, Sweden, Tunisia, Turkey, United Kingdom, USA and Zambia. This initiative has progressed from the COP26 Breakthrough Agenda to the COP27 Buildings Pavilion and officially launched at COP28.

The supporting initiatives are wide ranging and include the WBCSD built environment transformation, WGBC net zero-carbon buildings, WRI building efficiency accelerator, UNEP cool coalition, C40 building efficiency, IFC EDGE green buildings, OECD decarbonising buildings in cities and regions and Carbon Risk Real Estate Monitor (CRREM).

Among the various initiatives is the Buildings Breakthrough to push for near-zero emission and resilient buildings by 2030 which was launched by the Governments of Morocco and France with the UN Environment

Programme at COP28 (Dubai 2024). The work of the Buildings Breakthrough includes an annual assessment of progress closely aligned with the Global Status Report for Buildings and Construction a flagship publication of the UNEP and Global Alliance for Buildings and Construction. Priority international actions for 2024–2025 are:

1. Definitions for near-zero emission and resilient buildings (NZERB) and develop associated pathways for implementation.
2. Scale up low carbon procurement commitments and alliances, define technical requirements, share procurement best practices and policy to lead by example for government occupied buildings.
3. Increase availability of finance for NZERB with a focus on emerging and developing countries.
4. Identify knowledge gaps and share high-quality research and best practices.
5. Share capacity building resources and tools to include NZERB principles in buildings codes and education, training and accreditation programmes and frameworks.
6. Enhance international collaboration on NZERB (GlobalABC, 2024).

The most recent publication available on the GlobalABC database at the time of writing (March 2024) is tellingly entitled "Why are we still not adapting? Instigating adaptation in the built environment". It is an excellent report, the contents are summarised in Appendix 1 with remarks for consideration in future editions. The full report is worth reading in full.

A major omission found in the stakeholder groups is the role of banks and professional advisors. The built environment depends on debt financing and banks have a major role to play in influencing resilience and adaptation. Banks probably fall into the same category as insurance and reinsurance. Professional advisors include valuers, quantity surveyors, sales and letting agents, development consultants, tax advisors and others involved in working with property developers to deliver projects from conception to completion. This stakeholder group could be considered with architects and/or property and facility managers.

As considered in detail in Chapter 2, the business case for action (adaptation) is real; stakeholders have all the tools and technology they need but the pace of change is still too slow. Entrenched attitudes, little regulatory or financial incentive, or perhaps a lack of awareness of the cost of doing nothing. Even little willingness to understand the return on investment. According to Peter Dampier, Managing Director of Vision Zero Connect in Hong Kong, a typical 10,000 sq.m. office in Hong Kong uses over US$400,000 per year and could save 10% of that for a relatively modest investment in a Building Energy Management System (BEMS). Scale that up over a whole building portfolio and the savings are considerable. Office leases in Asia are typically around 3 years and this may explain the slow action and points towards the need for more landlord and tenant collaboration. We will return to this later.

All in for a Net Zero Built Environment Initiative is one of the initiatives supported by the World Resources Institute (WRI). WRI is a Washington DC-based organisation, founded in 1982, that has evolved into a global organisation with 1900 staff around the world and "works globally and in focus countries to meet people's essential needs; to protect and restore nature; and to stabilise the climate and build resilient communities" (World Resources Institute, n.d.).

We Mean Business Coalition (WMBC) is a US registered not for profit organisation that drives collaborative leadership to work with businesses to take action on climate change aligned with a 1.5 °C pathway. That target that means halving global emissions by 2030 (We Mean Business Coalition, n.d.). WMBC offers ideas for a 50% reduction in built environment-related CO2 emissions by 2030, specifically governments should:

1. Develop decarbonisation roadmaps to address whole life cycle emissions and encourage multi-stakeholder collaboration.
2. Introduce mandatory codes and standards to address whole life carbon and life cycle emissions.
3. Reduce fossil fuel use in 3% of existing buildings per year to 2030 by undertaking deep renovations.
4. Introduce data tracking systems and standards to monitor decarbonisation progress.
5. Use procurement standards to support decarbonisation along the value chain, build resilient, circular and resilient buildings and infrastructure.

6. Allocate public funding to R&D, innovation and talent development to support private sector actions (We Mean Business Coalition, n.d.).

The World Green Building Council (WGBC) comprises more than 80 Green Building Councils to be “the largest local-regional-global action network accelerating sustainability in the built environment” (WGBC, n.d.b). Working with government, businesses and organisations, WGBC supports the ambitions of both the UN Global Sustainable Development Goals and the ambitions of the Paris Agreement. A range of solutions are offered including collaboration for decarbonisation roadmaps, going beyond the business case, case studies of sustainable buildings, transition guides, a health and well-being framework and a showcase of actions taken by the Green Building Council network.

World Business Council for Sustainable Development (WBCSD), based in Geneva, was established in 1995 by a merger of the World Industry for the Environment and the Business Council for Sustainable Development following the success of the 1992 Rio Summit, in 1995. WBCSD helps businesses respond to sustainability challenges with a mission to “accelerate the transition to a sustainable world and to help make more sustainable businesses more successful” (WBCSD, n.d.). Together with WRI, WBCSD co-developed the Greenhouse Gas protocol, an early example of standardisation and accountability, in 1998.

To support built environment decarbonisation, WBCSD offers a building system carbon framework to help companies understand emissions in their value chain enabling developers, tenants and investors push for alignment in decarbonisation (WBCSD, 2020).

C40 Cities (C40) is a network of city mayors from around the world with a common goal to tackle the climate crisis. The organisation helps to raise climate ambition, influence the global agenda, build a movement and scale up climate action. Scaling up climate action includes adaptation and water, air quality, energy and buildings, food systems, transportation, ports and shipping, urban planning and design and waste management. In the energy and buildings sector, C40 “supports cities in implementing ambitious building energy efficiency initiatives and catalysing the cleaning up of the energy supply in line with a 1.5C climate mitigation target” (C40 Cities, n.d.).

In this respect networks and programmes include new building decarbonisation, private building decarbonisation, municipal building decarbonisation, clean energy, clean construction, China buildings and South Africa buildings. A range of accelerator guides, resources and policy frameworks for the sector are shared on the C40 website making a valuable resource available for all cities.

The governments of France and Germany initiated the Partnership for Energy Efficiency in Buildings (PEEB) at COP22 (Marrakesh 2016) with 3 key objectives:

1. Promote sustainable design and construction.
2. Minimise GHG emissions by cost effective reductions in energy demand.
3. Make use of untapped social, environmental and economic benefits (PEEB, n.d.).

The PEEB approach is to combine financing to fund energy efficiency in large projects with advisory support and expertise. Qualifying for financing comes with conditions for the project to meet. The programme includes PEEB Cool to provide support in 11 hot countries; Albania, Argentina, Costa Rica, Djibouti, Indonesia, Mexico, Morocco, Nigeria, North Macedonia, Sri Lanka and Tunisia. Other programmes are PEEB Med and PEEB ASEAN to support sustainable building initiatives in the Mediterranean region and Southeast Asia, respectively.

The timber building transformation is supported by Built by Nature, launched in 2021, who support professionals, academics and policymakers involved in the built environment to "decarbonise the built environment and protect nature". The approach taken is one of building coalitions and encouraging knowledge sharing. The founding partner of Built by Nature is the Laudes Foundation established in 2020 by the Brenninkmeijer family who set up Dutch retailer C&A. Other sustainability-related philanthropic work by the Foundation includes Fashion for Good established in 2017 and most recently the Labora Fund in 2022 to support labour rights in Brazil. The Laudes Foundation work in the built environment is focused on building-level performance, whole life cycle emissions (meaning circularity and not just energy), social impacts and geographically within Europe. Supporting Built by Nature is the means by which the Foundation helps accelerate timber buildings in Europe.

EIT Climate-KIC is a European agency, that supports countries, regions, cities and industries reach their climate goals through innovation and transformation. Created by the European Institute of Innovation and Technology (EIT) in 2010 as a climate change focused knowledge innovation community (KIC) it is a body of the European Union. Since 2010, according to its website it has worked in 60 countries involving over 400 partners on climate entrepreneurship, education and innovation to enable adaptation and mitigate climate change (EIT Climate, 2024).

A key programme of EIT Climate-KIC is NetZeroCities which advises the 112 cities selected by the European Commission in 2022 to be climate neutral by 2030. The project involves associations and networks, businesses, research organisations, think tanks and academia. Stated objectives are:

1. Establish an approach for cities to be climate neutral.
2. Support the means for cities to use innovation to promote systemic change.
3. Be a platform for services and expertise to help cities.
4. Develop the pipeline of cities working towards climate neutrality (EIT Climate KIC, n.d.).

The Net Zero Cities Capital Hub makes EIT Climate-KIC a stand out. It helps to facilitate public and private finance for city transitions through offering technical assistance and preparing the project for financing through diverse capital sources.

The Cool Coalition is a "global multi-stakeholder network that connects a wide range of key actors to facilitate knowledge exchange, advocacy and joint action towards a rapid global transition to efficient and climate-friendly cooling". Assembled by the UNEP in 2019 the Cool Coalition is based on the premise that "almost one-third of the world's population faces dangerous temperatures for more than 20 days a year, while heatwaves cause 12,000 deaths annually". Climate change means more cooling is needed, but man-made cooling drives electricity consumption. As urbanisation continues and fuels income growth, demand for cooling increases. Indoor cooling results in outdoor warming.

The cool coalition approach includes:

1. Improving planning and design to reduce the need for mechanical cooling.
2. Changing to renewable energy tariffs.
3. Improve the efficiency of air conditioning and refrigeration.
4. Protect the vulnerable from extreme heat.
5. Cooperation to achieve a full range of impactful benefits (Cool Coalition, n.d.a).

Pilot projects in Cambodia, India and Vietnam aim to promote passive cooling to mitigate urban heat islands, scaleable solutions to a whole-systems approach to urban cooling and Urban Cooling Action Plans. These incorporate nature-based solutions, urban design, district cooling, municipal incentives and sustainable public procurement and involve local cities, national ministries as well as the NGO support. As an example, the Vietnam project has 4 main parts:

1. Urban Cooling Action Plans to cover spatial analysis of extreme heat and cooling, the causes and policy, planning and finance implication.
2. A national cooling fund to provide financial support to manage cooling and extreme heat.
3. Sharing best practices of urban cooling to replicate successes from pilot projects.
4. Support for implementation of urban cooling under Nationally Determined Contributions and related policies (Cool Coalition, n.d.b).

International Code Council (ICC), established in 1994, provides model codes and building safety solutions to help deliver safe, sustainable and affordable buildings and communities (International Code Council, n.d.). The ICC has published a guide to decarbonising the built environment which considers whole life carbon, materials and processes and operational carbon and electrification.

The International Green Construction Code (IGCC) provides an up-to-date set of model code regulations to govern buildings and structures and is developed in collaboration with The American Institute of Architects (AIA), ASHRAE, The US Green Building Council (USGBC) and the Illuminating Engineering Council (IEC). The 2021 edition is broadly compatible with the 15 International Codes published by the International Code Council (ICC). IGCC includes site sustainability, water and energy efficiency, materials and resources, indoor environment and construction and operation plans. As the ICC states: "building development frequently converts land from biologically diverse natural habitat that helps manage rainwater to impervious hardscape with reduced biodiversity", the transfer from natural environment to built environment. IGCC and other standards seek to reduce the negative impact of this exchange without compromising the way that buildings serve the user in terms of indoor environment and financial performance (International Code Council, 2022).

The Building Resilience Index (BRI) provides a framework for hazard mapping and resilience assessment for building to help design, construct and retrofit buildings with embedded resilience and support adaptation of the built environment to climate change. BRI takes a web-based approach to identify, manage and disclose risks (wind, water, fire and seismic). It is an "innovation of the International Finance Corporation", part of the World Bank Group and supported by the Government of The Netherlands, the Australian Government and the Rockefeller Foundation (BRI, 2024).

Other industry associations represent professional groups, these include the Urban Land Institute (ULI), Building Owners and Managers Association (BOMA), Royal Institution of Chartered Surveyors (RICS), Royal Institute of British Architects (RIBA), America institute of Architects (AIA), CoreNet and a whole range of other international and local organisations that oversee and manage professional standards within the full range of functions that service the built environment. Others cover specialist areas, including sustainability such as the Institute of Environment and Assessment (IEMA), now Institute of Sustainability and Environmental Professionals (ISEP), or like the Chartered Association of Building Engineers (CABE) have sustainability courses and awareness as part of their Continuing Professional Development (CPD). A summary table of each of these and their respective public sustainability agendas is provided in Appendix 2.

To conclude this review of the institutions, industry associations and NGOs it is useful to compare the International Maritime Organisation (IMO), and GlobalABC. Bearing in mind that the built environment and shipping industries are very different and the IMO will be discussed in more detail in Chapter 11.

Global ABC has 290 members, including 42 national governments membership are national level industry associations, building material suppliers, professional bodies (including some of those listed above), academia and others. This means it is developing all the deep knowledge and research but it is not immediately or directly reaching the building owners, property developers, asset managers, banks and insurance companies. At the time of writing, Global ABC is closed for membership applications. IMO too focuses on membership at the State level and standard compliance is directed through International Conventions and the IMO regulatory framework.

Understanding the different histories of the two organisations is important. IMO is long established and has a long record of setting international regulations and gaining acceptance in response to the imperatives of maritime safety and navigation standards. At a global level the shipping industry has consolidated. In contrast, GlobalABC has a much shorter history and the number of property developers is greater and they are geographically dispersed. A simple internet search for "shipping companies" yields 161,000 results while "property developers" yields 2,490,000. With residential homeowners a stakeholder, it is worth considering the total number of homeowners worldwide, available data suggests that 28 countries have home ownership rates of over 80%, including China. The EU average is around 70% (World Population Review, n.d.). Attempting to regulate a large and diverse stakeholder group such as "worldwide property owners" through a single UN body, is of course, an impossible task and can only be left to individual governments at different levels.

The summary in Table 8.1 reveals differences in establishment, mission and sustainability agenda. Possibly the best that GlobalABC can hope to achieve is greater engagement with property developers and owners in rapidly urbanising regions. Bringing finance to the table to deliver sustainable new buildings, adoption of sustainable building codes where none exist, low cost and energy focused certification standards and support for retrofits in key aspects of existing building sustainability.

Table 8.1 Comparison of IMO and Global ABC website statements

	IMO	*Global ABC*
Introduction summary	"IMO is the global standard-setting authority for the safety, security and environmental performance of international shipping. Its main role is to create a regulatory framework for the shipping industry that is fair and effective, universally adopted and universally implemented. In other words, its role is to create a level playing-field so that ship operators cannot address their financial issues by simply cutting corners and compromising on safety, security and environmental performance. This approach also encourages innovation and efficiency. Shipping is a truly international industry, and it can only operate effectively if the regulations and standards are themselves agreed, adopted and implemented on an international basis. And IMO is the forum at which this process takes place".	"Founded at COP21, hosted by the United Nations Environment Programme and with over 330 members, including 42 countries, the Global Alliance for Buildings and Construction (GlobalABC) is the leading global platform for all buildings stakeholders committed to a common vision: A zero-emission, efficient and resilient buildings and construction sector".

	IMO	*Global ABC*
Key sustainability themes	"Shipping is an essential component of any programme for future sustainable economic growth. Through IMO, the Organization's Member States, civil society and the shipping industry are already working together to ensure a continued and strengthened contribution towards a green economy and growth in a sustainable manner. The promotion of sustainable shipping and sustainable maritime development is one of the major priorities of IMO in the coming years. As part of the United Nations family, IMO is actively working towards the 2030 Agenda for Sustainable Development and the associated SDGs. Indeed, most of the elements of the 2030 Agenda will only be realized with a sustainable transport sector supporting world trade and facilitating global economy. IMO's Technical Cooperation Committee has formally approved linkages between the Organization's technical assistance work and the SDGs. While the oceans goal, SDG 14, is central to IMO, aspects of the Organization's work can be linked to all individual SDGs. Energy efficiency, new technology and innovation, maritime education and training, maritime security, maritime traffic management and the development of the maritime infrastructure: the development and implementation, through IMO, of global standards covering these and other issues will underpin IMO's commitment to provide the institutional framework necessary for a green and sustainable global maritime transportation system" (IMO, n.d.).	"The GlobalABC: • **is a global advocate and a catalyst to action**: GlobalABC advocates for market transformation and focuses on catalysing action by defining a carbon neutrality strategy for the built environment; GlobalABC advocates the sector's importance for global climate action through international high-level events, including climate COPs and Buildings Pavilion; GlobalABC coordinates the implementation work of the Buildings Breakthrough and the Intergovernmental Council for Buildings and Climate. • **is a trusted platform to set targets and track progress**: GlobalABC tracks progress in its annual Global Status Report for Buildings and Construction, and its Global Building Climate Tracker, a new index to track progress in decarbonization in the sector; GlobalABC co-leads the Human Settlements Pathway under the Marrakech Partnership for Global Climate Action, which outlines the 2030 and 2050 decarbonization goals. • **supports countries in setting priorities and measures based on their situation**: GlobalABC develops policy guidance and global and regional buildings and construction roadmaps outlining aspirational targets, timelines, and key actions for essential policies and technologies, and, offering a model for national and city level buildings and construction roadmaps to support and raise the ambition of NDCs. • **supports the private sector transition with priorities and strategies** toward business models focused on decarbonising and increasing the resilience of buildings, including through its Market Transformation and Finance Hubs, and the BuildingToCOP Coalition" (GlobalABC, n.d.).

The Role of National Government

A publication by the World Green Building Council (WGBC) entitled "Beyond the Business Case. Why You Can't Afford Not to Invest in a Sustainable Built Environment" sets the case for transformation of the built environment in the context of the post-pandemic "new normal". The report sets out the drivers of the business case as policy at different levels of government, finance regulation such as carbon pricing and ESG reporting. Taken together, these give access to finance, corporate reputation, enhanced asset values, stronger investment case/reduced risk of stranded asset, operational efficiencies, reduced construction cost, improved circularity, lower insurance premiums and occupier preferences. Other social value drivers, according to the same report, include public health impact of COVID-19, private sector awareness, public sector policy and procurement preferences. These result in a case for well-being, jobs, resilience and equity benefits, supply chain enhancements in terms of safety, welfare and so on among construction workers (WGBC, n.d.a).

The drive towards net zero comes from national level, provincial and city level governments. It also comes from businesses, NGOs and pressure groups. Just 20 countries produce around 76% of global greenhouse gas emissions while more than 140 countries have set net zero targets.

National/Federal Government Targets

Commitments to cut emissions and adapt to climate change cascade down from the Nationally Determined Contribution (NDC) which set the targets at a national level. These were set under the Paris Agreement (PA) and are intended to be updated every 5 years. The plans set out how the targets are to be reached, how to monitor and verify progress and how the transformation will be financed. Each plan is meant to be based on data and science, they are meant to be ambitious, including transforming the way societies consume and produce as well as considering other aspects of sustainability including women's rights, indigenous communities and young people. In this way the NDCs can be linked to the SDGs providing a holistic and integrated approach. In this way, individuals, industry, food production, energy, transport and institutions are all considered (UN, n.d.), including, of course, the built environment. In the case of the built environment, the EU (climate neutral by 2050) is widely recognised as a standout for a comprehensive approach. In particular the Energy Performance of Buildings Directive many elements of which are covered in different chapters of this book; challenges around implementation abound.

The Kingdom of Bhutan is the world's only carbon negative country. This landlocked Himalayan country between China and India is roughly the size of Switzerland. 71% forested and 51% officially protected, the development philosophy of Bhutan is based on Gross National Happiness (based on governance, preservation and promotion of culture, sustainable development and environmental conservation) (Butler, 2025). A Business-as-Usual scenario will see this carbon neutral status come under pressure with expansion of the built environment (city development, industrial growth, hydropower and loss of forests). Nationally Determined Contribution 3.0 covers the period 2025–2027 and "enhances policy coherence, strengthens institutional capacity and promotes inclusive stakeholder engagement to help Bhutan remain carbon neutral and climate-resilience" (UNDP, 2025). Bhutan deserves recognition and serves as a reminder that emerging economies, facing economic development needs, cannot always simply transplant policy from more developed regions of the world. They have their own local aspirations, culture, expectations and challenges.

The Ministry of Ecology and Environment of the People's Republic of China published China's Policies and Actions for Addressing Climate Change (2022). This sets out China's progress, experience and approaches. The document addresses a strategy for combating climate change, mitigating climate change, adapting to climate change, improving policy systems and support and involvement in global governance related to climate change. The section entitled "accelerating the improvement of building energy efficiency" describes the energy conservation and green building efforts made in China. Including, use of solar and geothermal energy powered heating, photovoltaic power generation, increase in ultra-low and near-zero energy buildings, renovation of residential buildings and green construction of new buildings (Ministry of Ecology & Environment of the People's Republic of China, 2022).

Approved by President Joseph R. Biden Jr The United States of America Nationally Determined Contribution "Reducing Greenhouse Gases in the United State: A 2030 Emissions Target" (2021) includes a paragraph on buildings. "The emissions reduction pathways for buildings consider ongoing government support for energy efficiency and efficient electric heating induction stoves, and adoption of modern energy codes for new buildings. The United States will also invest in new technologies to reduce emissions associated with construction, including for high-performance electrified buildings" (The United States of America, 2021). Executive Order 14162 issued by President Donald J. Trump in January 2025 directed the US Ambassador to the United Nations to withdraw from the PA for the second time (previous withdrawal on 2020 after initially joining in 2016). 1 years notice for withdrawal is required (Haskett, 2025).

While the Chinese document is supported by statistics and detail, the American one reads more like a statement of intent. The Chinese document also refers to implementing a national strategy with reference to major regions and strategic areas all of which are inter-provincial while the American document makes no similar reference to State level or interstate implementation. On the other hand, the American document notes the approval of the President while the Chinese document is a Ministerial level publication that refers to Xi Jinping Thought on Promoting Ecological Civilization, it does not mention explicit approval.

Interesting as these documents are, the details for specific actions and the supporting legislation falls to the lawmakers of each country and each have their different governance systems. Some of the Chinese implementation at city level have already been touched on in Chapter 4 and some US building case studies were covered in Chapter 7.

Malaysia's Ministry of Natural Resources and Environmental Sustainability announced a new National Policy on Climate Change in September 2024. This updates the previous version released in 2009. The new version known as NPCC2.0 includes the principles of "common but differentiated responsibilities", a just and equitable transition, a whole society and nation approach and cross-sectoral and integrated solutions (Zaid Ibrahim & Co, 2024). Strategies include policy harmonisation, developing a low carbon economy, climate resilient industrial development, adoption of mitigation and adaptation measures, renewable energy and energy efficiency, increase resilience to climate change but policy, plans, programmes and projects, support knowledge-based decision-making, improve collaboration for implementation of climate change responses, increase community awareness and participation and strengthen involvement in international programmes (Climate Change Laws of the World, n.d.).

Singapore is a useful case study to consider in some detail. The city state falls under both National and City Level. Being just 200 kilometres north of the equator it is subject to more intense sunlight and higher temperatures than other latitudes which can result in more potent pollutants and the effect of heat islands and need for cooling is intense. It also results in stronger thermal currents delivering greenhouse gases higher into the atmosphere (Benos, 2016).

The Singapore Green Plan 2030 (SGP) is notable for the inter-agency approach being jointly prepared by Ministries of Education, National Development, Sustainability and the Environment, Trade and Industry and Transport. The Singapore Green Building Masterplan (SGBMP) is the action plan for the built environment and was first launched in 2006. The current version for it aims to "foster a leading green built environment sector that can lower its carbon footprint and provide a healthy, liveable and sustainable built environment for all". The 80-80-80 goals refer to 80% of buildings to be green by 2030, 80% of new developments to be Super Low Energy (SLE) by 2030 and achieving an 80% improvement in energy efficiency for the best green buildings by 2030 (Ministry of National Development, 2021). The Building and Construction Authority (BCA) administers the Singaporean BCA Green Mark certification, which is tailored for the demands of the local climate. The 2021 update is aligned with the SDGs as well as improving energy efficiency standards and an increased emphasis on maintainability, reducing life cycle embodied carbon, improving resilience and healthier environments for building users (BCA, n.d.)

Having a comprehensive green building master plan sets Singapore apart from many other nations. Even more notable is the specific inclusion of existing buildings and tenant engagement. All of which, together with the SLE and public engagement feedback is incorporated into the SGP.

The aims of the SGBPM are summarised as follows:

1. 80% of building GFA green certified by 2030.
2. BCA to identify buildings where energy performance data is published.
3. Raise energy performance requirements and sustainable construction in new buildings and those undergoing major retrofits.
4. Update BCA to raise energy performance requirements, emphasise design, embodied carbon and healthier environments for users.

5. Introduction of Green Mark Incentive Scheme for Existing Buildings 2.0 to help increase energy efficiency in existing buildings. S$63 million budget available until the sooner of funds committed or 31st March 2027.
6. 80% of new buildings by GFA to be SLE from 2030.
7. Mainstreaming SLE by driving SLE standards for existing buildings undergoing deep retrofit where feasible.
8. New SLE standard for residential buildings where 60% improvement achieved compared to 2005 levels.
9. 3% GFA bonus for Green Mark SLE with Maintainability Badge and other targets.
10. Push for improvement in energy efficiency from 70% to 80% through the Green Buildings Innovation Cluster (GBIC) programme supported with S$45 million funding (BCA, 2024).

Singapore also makes use of a Growth and Transformation Scheme (GTS) to encourage strategic alliances between construction, developers and professional advisors targeting value chain transformation. This includes use of technology at the design and delivery stage to drive efficiency. Established in 2010 the Construction Productivity and Capability Fund (CPCF) uses financial incentives to help adoption of technology. Adoption of labour efficient designs and technology are supported by the Buildability Framework (BCA, 2021).

While this section is not intended to be a comprehensive piece of research across all national plans, Singapore is generally considered a leader in implementing building sustainability. The use of industry alliances and clear governance frameworks and targets together with publication of data and cash incentives provide a case study for others to learn from. Singapore also has a carbon tax first established under the Carbon Pricing Act (CPA) in January 2019 and amended in late 2022 (National Environment Agency, 2024). While the tax applies only to large upstream emitters, including power stations, petrochemicals, oil refineries and similar, these account for around 79% of Singapore emissions (National Climate Change Secretariat, n.d.). Having this on the statute book must also be a clear signal for all industries to take action.

Canada is a "federal bicameral parliamentary democracy" and has local government legislation to give "powers of general competence" that vary at the provincial and territorial level (Commonwealth Local Government Forum, n.d.). British Columbia and Vancouver City are considered separately in the Local Government section below. At the federal government level, the Canada Green Buildings Strategy (CGBS) presents the national plan for energy efficient and affordable buildings and homes. It includes making buildings more resilient and retrofits for homes and commercial buildings. ICGBS is a cross Ministry document co-signed by the Ministries of Energy and Natural Resources, Housing, Infrastructure and Communities and Environment and Climate Change and aligned with the Canadian Net-Zero Emissions Accountability Act. For commercial buildings, there is a "deep retrofit accelerator initiative and budgets for implementing buildings standards, labelling, benchmarking and disclosure" (Wilkinson, 2024).

Pakistan

The NDC Implementation Plans for Pakistan includes a conditional target to reduce emissions by 50% by 2030. Various supporting plans cover different industries and priority actions (UNDP, 2024).

There are 3 comprehensive documents covering green buildings and building efficiency in Pakistan. These are:

1. Vision 2030 for a Green Building Code in Pakistan, 2022.
2. Green Building Code of Pakistan, 2023.
3. Energy Conservation Building Code—2023.

Pakistan Vision 2030 includes a high-level strategic framework for sustainable development, with the overarching aim of embracing sustainability to transform Pakistan into a prosperous, industrialised, and equitable society by 2030. Key objectives include:

1. Economic Growth through knowledge and resource optimisation.
2. Reducing poverty and inequality while ensuring inclusive growth and access to basic services for all citizens.
3. Environmental Sustainability by promoting green practices, renewable energy and conservation.
4. Education and Innovation through education, skill development and technological advancements.

5. Infrastructure Development to modernising transportation, energy and urban infrastructure to support economic activities and improve living standards.

When it comes to the built environment specifically it aims to reduce greenhouse gas emissions, improve resource efficiency and create a sustainable urban environment by:

6. Developing and implementing a comprehensive Green Building Code to focus on energy efficiency, water conservation, and the use of sustainable materials throughout a building's life cycle.
7. Adoption of energy efficient designs, renewable energy sources and technologies like solar panels and efficient HVAC systems.
8. Introducing rainwater harvesting, wastewater recycling and efficient water use systems in buildings.
9. Plans to retrofit older buildings to improve energy efficiency and reduce their environmental impact.
10. Training architects, engineers and construction professionals in green building practices and integrating these concepts into educational programs.
11. Launching pilot projects, particularly in public buildings, to demonstrate the benefits of green construction and create market demand.
12. Offering financial incentives, such as tax rebates and low-interest loans, to encourage the adoption of green building practices.
13. Designing buildings to withstand climate challenges like extreme temperatures, flooding and earthquakes (SCP Facility, 2022).

The Green Building Code of Pakistan—2023 is based on the International Code Council's International Code (International Code Council, 2023). It includes specific provisions for embodied carbon as follows:

14. Low carbon and locally sourced materials to minimise the embodied carbon associated with transportation and production.
15. Life cycle assessments (LCA) of building materials to evaluate and reduce their carbon footprint throughout their life cycle.
16. Reuse and recycling of construction materials.

17. Innovative construction techniques and materials, such as low carbon and carbon negative concrete.
18. Mandatory reporting and mitigation of embodied carbon in large-scale projects, particularly in urban areas (PEC, 2023).

The Gazette of Pakistan published the Energy Conservation Building Code 2023 (ECBC-2023), on 15th March 2024, for immediate implementation and to apply to all new and old buildings. The National Energy Efficiency and Conservation Authority (NEECA) are developing a building rating system to harmonise ECBC-2023 with LEED, incorporate it into local building regulations and facilitate certified energy auditors and managers use of the code in retrofits. It is difficult to overstate how comprehensive this 190 page mandatory code is. Table 8.2 directly summarises the section summaries in order to share the leadership and commitment to energy conservation shown by the Government of Pakistan, Ministry of Energy.

Local Government, Cities, City Districts and Local Councils

Finding state or provincial policies or initiatives for decarbonising the built environment in Asia Pacific can be difficult. The focus is very much on decarbonising energy, transportation as well as mitigation and adaptation efforts. Perhaps existing building codes and other regulations obviate the need for a comprehensive building decarbonisation strategy.

At the sub-national government level, Victoria State Government in Australia has published "Victoria's 2035 Emissions Reduction Target". It sets out initiatives and plans covering renewable energy, helping households transition, transport emissions, industry transition, agriculture and environmental improvement. Other than solar homes, there is no specific strategy for commercial property and the built environment.

As has been hinted at in the Singapore case above, things seem to get more interesting at city level. According to the Carbon Disclosure Project (CDP), 41% of cities in the Asia Pacific region had set an emissions reduction target by 2022 with a further 10% planning to introduce one by the end of 2024. The same source suggests that building mitigation only accounts for 5% of actions being taken at city level in APAC (CDP, n.d.). This represents a significant gap in emissions reduction plans.

The Asia Business Council, founded in the wake of the 1999 Asian Financial Crisis is a body that aims to increase dialogue between Asian

Table 8.2 Summary of Pakistan building energy conservation code

Section	*Summary/Scope*
Introduction	• Building sector consumes 23% of energy and 54% of electricity in Pakistan. • ECBC-2023 aims to improve energy efficiency. • Code is to be revised every 3 years.
Administration and enforcement	• Provision for an administrative framework and sets responsibilities for stakeholders. • Enforcement relies on co-ordination, training and capacity building. • Provision for Statutory Regulatory Order to mandate demonstration of energy efficient designs.
Urbanisation and developments	• Sustainable, compact urban development. • Energy efficiency, green spaces and efficient transport.
Building envelope	• Plays a vital role in ensuring energy efficiency and sustainability. • Also maintains air quality and comfort.
Passive building design	• Tailored for hot and humid climates. • Minimise heat gain. • Enhance natural cooling.
Retrofitting of existing buildings	• Regulatory frameworks. • Incentives. • Stakeholder engagement. • Implementation.
Building insulation	• Minimum requirements. • Evaluating effectiveness. • Maintenance.
Heating, ventilation and air conditioning	• Minimum energy efficiency standards. • Optimisation of HVAC systems. • Covers equipment, installation, operation, maintenance, ductwork, piping, insulation and controls.
Water heating	• Design, installation, operation and maintenance of systems. • Save energy, lower costs and reduce emissions.
Lighting	• Minimum efficiency standards, lighting requirements, controls and design.

(continued)

Table 8.2 (continued)

Section	*Summary/Scope*
Electricity	• Evaluate load, select power sources, distribution sources. • Consider transformers, circuit breakers and safety standards.
Monitoring	• Use of devices to monitor energy inputs and outputs. • Retain historic data for future reference. • Establish benchmarks.
Geothermal energy	• Source of renewable energy.
Renewable energy	• Sources, systems, installation. • Industry codes and standards for safety.
Energy management system	• Energy policies, processes and objectives. • Hardware and software requirements. • Target setting and implementation. • Continuous improvement.
EV charging	• Design requirements for electricity supply and ventilation. • Safety requirements. • Provision for future installation.
Water recycling	• Treatment and disinfection of wastewater for non-drinking reuse.

(NEECA, 2024)

and international business leaders, support the role of private business in economic development and growth and be an influence on government, NGOs, business and civil society. The publication "Cities: Asia's Catalysts for Net Zero" highlights the role of cities in Asia as exemplars of decarbonisation given their roles as sources of human and financial capital, innovation centres, policy led solutions and the high density of people and activity to yield efficiency of resource and solution scaleability. The opportunities include:

1. Policy experimentation, including emissions trading.
2. Climate resilience, including flood defences, biodiversity, green buildings and water and food security, compact urban models, low carbon built environment, green and blue infrastructure, carbon sinks, circularity.

3. Reallocation of capital, including green finance, carbon pricing, incentives such as capital requirements and blended finance.
4. Green innovation, including access to funding, policy support, nurturing talent, research centres, interactions between finance, research and industry and sustainability awareness (Voets et al., 2023).

One city that is taking adaptation action is Jakarta. It is moving. Or at least the administrative capital of Indonesia is moving to Nusantara in Borneo. The new capital has a projected population of nearly 2 million by 2045 while 95% of North Jakarta is expected to be lost to the sea by 2050, with 1.8 million people affected (Voets et al., 2023). Relocating the administrative functions of a city still leaves the existing urban areas at risk, albeit in different degrees. Rather than depend on traditional regional analysis, Nasutin et al (2022) considered local variations in vulnerability to natural hazards within Jakarta at the local level by using fuzzy geographically weighted clustering and chaotic flower pollination algorithm to improve the results. The data-driven approach reveals flood risk in sub-districts and the most vulnerable times of the year (Nasution et al., 2022).

The Kuala Lumpur (KL) "Low Carbon Society Blueprint 2030" 4th Edition is dated July 2018, and like many cities, the efforts to decarbonise are based on a transition to low carbon grid energy and energy efficiency (C40 Cities, 2021). Under the section for Low Carbon Green Building, Commercial Buildings, the document envisages a transition between 2020 and 2030 to high efficiency air conditioning, conventional electric water heaters to high efficiency versions and solar, a significant increase in solar power generation (40% of buildings by 2030) and insulation and behavioural energy savings. The report tagline "70 by 30" refers to a target reduction of carbon emission intensity of GDP by up to 70% by 2030. Of this just over 20% is to come from low carbon green building and nearly 34% from a sustainable energy system. The detailed plan for green buildings includes programmes under the headings: building envelope, mitigation of heat island, water efficiency, low carbon construction, energy efficient air conditioning, lighting and other systems, green building monitor, green building plan, sustainable management and audit and global warming control substances. It is a thorough and impressive document setting out the framework and required actions (City Planning Department, 2018).

In New York, Local Law 97 (LL97) is part of the Climate Mobilisation Act aligned with 2050 carbon neutrality. It aims to address the nearly 70% of carbon emissions from fossil fuels used to heat, cool and power buildings (NYC Accelerator, n.d.). It applies to buildings of over 25,000 square feet (2322 sq.m) who must input their energy usage over the year. Penalties are imposed for failing to meet standards with increasing requirements over a period of time. According to NYC Accelerator, in 2022, 63% of buildings exceed limits for the 2030 standard. The major criticism of the initial scheme is that while it encourages elimination of waste, it failed to account for electrification being a key step towards decarbonisation and net zero readiness (Kaufman & Tan, 2020). When a fully electric building switches to a renewable tariff that a major step towards net zero building operations.

The province of British Columbia Local Government Act gives power to local governments to develop community plans. The City of Vancouver has different arrangements and community plans fall under the Vancouver Charter and these may be given different titles in different Local Planning Areas. Among other things, community plans in British Columbia and Vancouver City seek to promote energy efficiency to address climate change and resource depletion (City of Vancouver, 2011). The British Columbia Climate Action Charter commits almost all local government to work towards carbon neutral emissions at the operating level, measure and report on community emissions and create energy efficient communities (Government of British Columbia, 2023).

The requirements for energy efficiency in residential houses are covered under Section 9.36 of the British Columbia Building Code. This covers the building envelope, water heaters and air conditioning. Not least is the requirement for a heat recovery ventilation system or dehumidification for spaces with an indoor pool or hot tub (Higgins & Finch, n.d.)

Summary

As we have seen from this chapter, the global industry body for the built environment exists but it lacks the history and participation of some others. Sustainability in the built environment depends on the success of GlobalABC adopting some of the IMO capacity to enable it to shape national and city government regulations to push and pull all stakeholders to take action. We will see in Chapter 11 that other industries have had successes, most notably a multinational industry body, cross-border cooperation on harmonisation of standards, a business case for process transformation, technological innovation and market leaders of sufficient scale to influence the industry as a whole.

Even to the trained eye, the range of organisations, government targets and policy frameworks, initiatives, is overwhelming. This reflects the complexity of the industry but there seems to be a greater role for national policy framework making and target setting at the city or municipality level. Simplifying and standardising metrics and publishing data would be a good start. If city regulations included publishing building energy consumption that would enable occupiers to make energy efficiency-related choices. Setting green certifications aside, understanding building energy use intensity and renewable power generation on the operations side would be a start towards retrofitting to improve energy performance or our sample case non-core CBD office tower in Hong Kong. Recognising, financing and mainstreaming low carbon materials on the embodied carbon would encourage changes in material selection and updating building codes can also help.

The role of government, from Bhutan to China and Singapore to the USA, as a driver of change includes national level policy frameworks and funding and local level implementation and regulations. Almost all governments recognise the need to make changes, not just in energy efficiency but across the built environment value chain. There are different levels of activity required. The role of GlobalABC, in collaboration with all the industry organisations, is to act as a bridge for continuous support and expertise from the leading national and city governments to those at an earlier stage. A particular focus might be city leader engagement, this is the level where implementation policy is made. Building sustainability, particularly energy efficiency is a central means of meeting NDCs, there is no hiding from that. Getting the co-ordination, collaboration and governance model right is essential to avoid increases in built environment-related emissions in rapidly urbanising economies everywhere.

References

Benos, T. (2016, November 8). *Pollution emitted near equator has biggest impact on global ozone*. Science Daily. Retrieved October 14, 2024, from https://www.sciencedaily.com/releases/2016/11/161108085028.htm

Building and Construction Authority. (2021, March 4). *Push for greener buildings and new support scheme to boost transformation efforts through industry alliances unveiled*. Retrieved 2024, from https://www1.bca.gov.sg/docs/default-source/docs-corp-news-and-publications/media-releases/bca-cos-2021-media-release.pdf

Building and Construction Authority. (2024, July 18). *Green building masterplans*. Building and Construction Authority (BCA). Retrieved November 14, 2024, from https://www1.bca.gov.sg/buildsg/sustainability/green-building-masterplans

Building Resilience Index. (2024). *About — Building resilience index*. Building Resilience Index. Retrieved November 14, 2024, from https://www.resilienceindex.org/about

Building and Construction Authority. (n.d.). *Sustainability - Green Mark Certification Scheme - Green Mark 2021*. Building and Construction Authority (BCA). Retrieved November 14, 2024, from https://www1.bca.gov.sg/buildsg/sustainability/green-mark-certification-scheme/green-mark-2021

Butler, S. (2025, August). A kingdom built on happiness. *Geographical*, 31–38.

C40 Cities. (2021, October). *Kuala Lumpur as a climate-smart and low carbon city*. C40 Cities. Retrieved November 14, 2024, from https://www.c40.org/case-studies/kuala-lumpur-as-a-climate-smart-and-low-carbon-city/

C40 Cities. (n.d.). *Energy & buildings*. C40 Cities. Retrieved November 14, 2024, from https://www.c40.org/what-we-do/scaling-up-climate-action/energy-and-buildings/

CDP. (n.d.). *Asia Pacific cities climate finance snapshot*. CDP. Retrieved November 14, 2024, from https://www.cdp.net/en/research/global-reports/asia-pacific-cities-climate-finance-snapshot#notes

City of Vancouver. (2011, July 28). *Vancouver's next community plans*. Retrieved 2024, from https://vancouver.ca/files/cov/next-community-plans.pdf

City Planning Department. (2018, July). *Kuala Lumpur low carbon society blueprint 2030*.

Climate Change Laws of the World. (n.d.). *National policy on climate change*. Retrieved November 14, 2024, from https://climate-laws.org/documents/national-policy-on-climate-change_b509?id=national-policy-on-climate-change_4808

Commonwealth Local Government Forum. (n.d.). *The local government system in Canada*. Retrieved 2024, from https://www.clgf.org.uk/default/assets/File/Country_profiles/Canada.pdf

Cool Coalition. (n.d.a). *Overview*. Cool coalition. Retrieved November 14, 2024, from https://coolcoalition.org/about/overview-and-history/

Cool Coalition. (n.d.b). *Sustainable urban cooling in Viet Nam*. Cool Coalition. Retrieved November 14, 2024, from https://coolcoalition.org/pilot-projects/sustainable-urban-cooling-in-viet-nam-cities/

Cushman & Wakefield. (n.d.). *Built environment: Going beyond net zero to do more good*. Cushman & Wakefield. Retrieved November 14, 2024, from https://www.cushmanwakefield.com/en/insights/build-environment-going-beyond-net-zero-to-ng-more-good

EIT Climate. (2024). *About EIT climate-KIC*. Climate-KIC. Retrieved November 14, 2024, from https://www.climate-kic.org/who-we-are/about-eit-climate-kic/

EIT Climate KIC. (n.d.). *About net zero cities*. Climate-KIC. Retrieved November 14, 2024, from https://www.climate-kic.org/spotlight-initiatives/netzerocities/

GlobalABC. (2021, 11 2). *Market transformation levers for a net zero built environment*. GlobalABC. Retrieved November 14, 2024, from https://globalabc.org/resources/publications/market-transformation-levers-net-zero-built-environment

GlobalABC. (2024, March 8). *Buildings breakthrough priority international actions for 2024-2025 launched at the buildings and climate global forum*. GlobalABC. Retrieved November 14, 2024, from https://globalabc.org/news/buildings-breakthrough-priority-international-actions-2024-2025-launched-buildings-and-climate

GlobalABC. (n.d.). *Mission and vision*. Globalabc. Retrieved March 5, 2024, from https://globalabc.org/about/mission-and-vision

Goodier, C. (n.d.). *The urgent need for a net zero built environment*. Loughborough University. Retrieved October 14, 2024, from https://volume.lboro.ac.uk/net-zero-built-environment/

Government of British Columbia. (2023, January 9). *Local government planning for sustainability and resilience - Province of British Columbia*. Gov.bc.ca. Retrieved November 14, 2024, from https://www2.gov.bc.ca/gov/content/governments/local-governments/planning-land-use/local-government-planning/planning-for-sustainability-resilience

Haskett, J. D. (2025, April 14). *U.S. Withdrawal from the Paris Agreement: Process and potential effects*. Congress.gov. Retrieved June 22, 2025, from https://www.congress.gov/crs-product/R48504

Higgins, J., & Finch, G. (n.d.). *Energy efficiency requirements for houses in British Columbia*. Retrieved 2024, from https://www2.gov.bc.ca/assets/gov/farming-natural-resources-and-industry/construction-industry/building-codes-and-standards/guides/climatezone5-7a.pdf

International Code Council. (2022, April). *Preface - 2021 International Green Construction Code (IGCC)*. ICC Digital Codes. Retrieved November 14, 2024, from https://codes.iccsafe.org/content/IGCC2021P2/preface

International Code Council. (n.d.). *Who are we*. ICC - International Code Council - ICC. Retrieved November 14, 2024, from https://www.iccsafe.org/

Kaufman, N., & Tan, Y. A. (2020, October). *New York City's building emissions law shows the importance of economy wide climate policy.* Retrieved 2024, from https://www.energypolicy.columbia.edu/sites/default/files/file-uploads/NYC_Buildings_CGEP_Commentary_FINAL_r1.pdf

Ministry of National Development. (2021, April 28). *MND | Singapore Green Building Masterplan.* Ministry of National Development (MND). Retrieved November 14, 2024, from https://www.mnd.gov.sg/our-work/greening-our-home/singapore-green-building-masterplan

Ministry of Ecology and Environment of the People's Republic of China. (2022). *China's policies and actions for addressing climate change.* Ministry of Ecology and Environment. Retrieved November 14, 2024, from https://english.mee.gov.cn/Resources/Reports/reports/202211/P020221110605466439270.pdf

Nasution, B. I., Saputra, F. M., Kurniawan, R., Ridwan, A. N., Fudholi, A., & Sumargo, B. (2022). Urban vulnerability to floods investigation in jakarta, Indonesia: A hybrid optimized fuzzy spatial clustering and news media analysis approach. *International Journal of Disaster Risk Reduction, 83.* https://doi.org/10.1016/j.ijdrr.2022.103407

National Environment Agency. (2024, August 12). *Carbon Tax - Singapore.* The National Environment Agency. Retrieved November 14, 2024, from https://www.nea.gov.sg/our-services/climate-change-energy-efficiency/climate-change/carbon-tax

National Climate Change Secretariat. (n.d.). Frequently asked questions. Retrieved November 14, 2024, from https://www.nccs.gov.sg/files/docs/default-source/default-document-library/annex-b.pdf

NEECA. (2024). *Energy conservation building code 2023.* https://neeca.gov.pk/SiteImage/Downloads/ecbc23.pdf

NYC Accelerator. (n.d.). *Local Law 97.* NYC Accelerator. Retrieved November 14, 2024, from https://accelerator.nyc/ll97

Partnership for Energy Efficiency in Buildings. (n.d.). *About PEEB.* PEEB – Transforming the building sector. Retrieved November 14, 2024, from https://www.peeb.build/about-peeb

PEC. (2023). *Green building code of Pakistan - 2023.* Retrieved April 25, 2025, from https://www.pec.org.pk/

Reynolds, M. (2024, March 19). *How can bold leadership propel the built environment towards net zero?* Mace Group. Retrieved November 14, 2024, from https://www.macegroup.com/perspectives/articles/2024/march/how-can-bold-leadership-propel-the-built-environment-towards-net-zero

SCP Facility. (2022, March). *Vision 2030 for a green building code in Pakistan.* Retrieved April 25, 2025, from https://www.switch-asia.eu/site/assets/files/3366/pakistan_vision_2030_final.pdf

The United States of America. (2021, April 15). *The United States of America: Nationally determined contribution*. UNFCCC. Retrieved November 14, 2024, from https://unfccc.int/sites/default/files/NDC/2022-06/United%20States%20NDC%20April%2021%202021%20Final.pdf

The Breakthrough Agenda. (n.d.). *About the breakthrough agenda*. Breakthrough Agenda. Retrieved November 14, 2024, from https://breakthroughagenda.org/about/

UN. (n.d.). All about the NDCs | United Nations. the United Nations. Retrieved November 14, 2024, from https://www.un.org/en/climatechange/all-about-ndcs

UNDP. (2024, November 19). *NDC investment plan Pakistan*. United Nations Development Programme. Retrieved June 22, 2025, from https://www.undp.org/pakistan/publications/ndc-investment-plan-pakistan

UNDP. (2025, June 4). *Bhutan's nationally determined contributions 3.0*. United Nations Development Programme. Retrieved August 15, 2025, from https://www.undp.org/bhutan/publications/bhutans-nationally-determined-contributions-30

United Nations. (n.d.). *Net Zero coalition | United Nations*. the United Nations. Retrieved November 14, 2024, from https://www.un.org/en/climatechange/net-zero-coalition

Voets, B., Howe, C., & Pau, J. (2023, March). *Cities: Asia's catalysts for net zero*. Asia Business Council.

WBCSD. (n.d.). *Our history*. The World Business Council for Sustainable Development (WBCSD). Retrieved June 22, 2025, from https://www.wbcsd.org/who-we-are/our-history/

We Mean Business Coalition. (n.d.). *We are a global nonprofit Coalition working with the world's most influential businesses to take action on climate change*. We Mean Business Coalition. Retrieved November 14, 2024, from https://www.wemeanbusinesscoalition.org/about/

WGBC. (n.d.a). *Beyond the business case*. World Green Building Council. Retrieved November 14, 2024, from https://worldgbc.org/wp-content/uploads/2022/08/WorldGBC-Beyond-the-Business-Case.pdf

WGBC. (n.d.b). *Who we are*. World Green Building Council: Home. Retrieved November 14, 2024, from https://worldgbc.org/

Wilkinson, J. (2024, August 16). *Departmental strategies*. The Canada green buildings strategy: Transforming Canada's buildings sector for a net-zero and resilient future. Retrieved November 14, 2024, from https://natural-resources.canada.ca/transparency/reporting-and-accountability/plans-and-performance-reports/departmental-strategies/the-canada-green-buildings-strategy-transforming-canadas-buildings-sector-for-net-zer/26065

Wilmot, B. (2024, May 10). *There is no future for buildings that are not green, says JLL boss Christian Ulbrich*. Real Commercial. Retrieved November

14, 2024, from https://www.realcommercial.com.au/news/there-is-no-future-for-buildings-that-are-not-green-says-jll-boss-christian-ulbrich

World Resources Institute. (n.d.). *About WRI*. World Resources Institute. Retrieved November 14, 2024, from https://www.wri.org/about

World Population Review. (n.d.). *Home ownership by country 2024*. World Population Review. Retrieved November 14, 2024, from https://worldpopulationreview.com/country-rankings/home-ownership-by-country

Zaid Ibrahim & Co. (2024, October). *Decoding Malaysia's climate agenda: Key takeaways from the NPCC2.0 and Climate Change Bill*. Retrieved November 14, 2024, from https://www.ziclegal.com/insights/decoding-malaysias-climate-agenda-key-takeaways-from-the-npcc2-0-and-climate-change-bill

CHAPTER 9

Green Building Certifications

"We often think of buildings as inert structures, but they are living systems that either enable or erode our ability to thrive. By understanding how to design with climate health in mind, leaders can become stewards or environments that foster true flourishing." (Professor Khee Poh Lam, Provost's Chair Professor of Architecture and the Built Environment, National University of Singapore).

What Do Green Building Certifications Do?

Having looked extensively at the sources of carbon emissions across the built environment, this chapter examines one of the key tools available for property developers and building owners. Green certifications apply to building design (or new construction) and operations (or in-use), this distinction may not be widely known or understood. The design certification is based on meeting criteria for sustainable design while the in use or operation certification is based on actual sustainability performance. The latter is probably of more relevance for achieving net zero targets but it is difficult to know if a new building is on course without a sustainably designed project in the first place. For the purposes of establishing green credentials and new building marketing, the design certifications may be considered sufficient.

S. Crispin, *Sustainability in the Built Environment*, https://doi.org/10.1007/978-981-95-6549-8_9

Green building certifications aim to encourage the building industry to:

1. Reduce the harmful effects of the built environment on the natural environment;
2. Cut GHG and use of fossil fuels, and;
3. Improve the quality of life for those who live and work in the buildings.

Andy To is Managing Director, North Asia of US Green Building Council and he emphasises item 3 in the list above: "the user experience is part of the objective". By extension, and as discussed previously, this is very much part of building competitiveness and a key part of the business case.

The stakes are very high. It is worth repeating that the built environment is a significant contributor to global GHG emissions: 40% is the commonly quoted figure but perhaps more reliably "around 37% of energy and process related CO2 emissions", 30% of raw materials, 25% of water and 12% of land and at the end of life 25% of solid waste (UNEP, 2009). Anything that helps property developers and owners measure and manage their GHG emissions can only be a good thing environmentally and from a business point of view too.

Pros and Cons of Green Building Certifications

For property developers, there are pros and cons to green certifications. On the plus side, it is a mark of official and societal approval of their buildings and may qualify for tax incentives, grants, green finance or other rewards. In Hong Kong and Singapore, green buildings can benefit from a Gross Floor Area (GFA) concession which can be up to 10% of the total GFA in Hong Kong (Chan & Leung, 2017). Reduced energy consumption means lower costs and other longer term operational efficiency. Green certification makes buildings more attractive to office tenants who may list green certification as a "search requirement" and their employees who want to see the company they work for implementing sustainability into their operations resulting in talent retention. In particular, the use of sustainable materials in office fit outs can help overcome employee reluctance to take up seats in a newly renovated office.

As discussed in Chapter 2, the "green premium" for office space in Asia varies quite widely. According to Savills Research & Consultancy, from a little over 2% in Tokyo to around 18% in Bangkok but the greenest buildings are also generally the newest which always attract a premium and that is likely to skew the numbers (Savills, 2022). In London, A new net zero office building achieved rents that were higher than expected, it is rated BREEAM Outstanding which means it is in the top 1% of new non-residential buildings (Specification Online, 2023). In reality, newer buildings tend to be green from the ground up and new buildings also tend to attract a premium. It is difficult to separate the premium due to being green from that of being new.

At the same time there is also a brown discount and the risk of assets becoming "stranded", or so unattractive to the market that they are unable to be let in their current condition, but they also cannot secure a rent that is high enough to make the investment worth it' (Oliver, 2023). Inevitably this has an impact on valuations. The case of 20 Canada Square in London's Canary Wharf offers some insights, according to media reports the building was bought for GBP410 million in 2014 and was reportedly listed for sale at GBP250 million in March 2023. In the intervening period, a key tenant moved out and the property was reported to be in need of significant investment to upgrade energy efficiency to attract new tenants (The Business Times, 2023). There are of course many other factors at play, including the trend towards hybrid working, economic slowdown, high interest rates, high energy costs, oversupply in some markets and so on.

Conversely, and as discussed in Chapter 2, there is more and more evidence that the greener the building the more likely it is to attract a higher valuation according to a report published by the Royal Institution of Chartered Surveyors (Day, 2022). It is also a positive for investors who consider ESG an important investment screening criteria. ESG property funds will target more sustainable and green-certified buildings and organisations like the Net Zero Asset Owner Alliance, a United Nations convened initiative of institutional investors each committed to transitioning their portfolio to net zero greenhouse gas emissions by 2050 (UNEPFI, n.d.) and the Net Zero Asset Managers initiative with similar goals (Net Zero Asset Managers, n.d.).

In summary, common benefits of green building certification include:

1. Operational efficiency

2. Qualification for tax incentives
3. Meeting regulatory requirements
4. Attracting talent and retention
5. Tenant satisfaction
6. Energy conservation
7. Water conservation
8. Responsible use of materials
9. Project management team time savings
10. Future proofing asset value
11. Increased occupancy
12. Accountable contractors
13. Industry recognition (Green Building Initiative, n.d.; CIM, n.d.).

May Lv works for Savills in Guangzhou, with multiple green building assessment qualifications including LEED Fellow, CFA-ESG and WELL AP. She sees the wide adoption of green building certification in China as part of a transformation of the real estate industry from one of short-term profit maximisation to a long-term sustainable income model. With government support, including subsidies for ultra-low energy buildings as well as tenant and occupier demand, green certification has become part of a virtuous cycle for investors to avoid policy risks, achieve rent premiums, increase occupancy rates and attract high-quality tenants.

On the downside, the conditions required to obtain a certification will increase the cost of the building and could extend the construction period. According to the Hong Kong Polytechnic University & Construction Industry Institute, the cost of the certificates vary up to 4% of a new office building, depending on the level of certification sought. May Lv says the additional cost may be 10–25% that of a traditional building but this can be offset by benefits mentioned above as well as future proofing against possible carbon tax, higher renovation costs and access to green finance.

In addition, obtaining the certification takes time and involves complicated procedures, although the right support can streamline the process if planned in advance. Some certifications require two assessments and some are based on data collected over a period of time after the building is operational. In some cases, the uncertainty of the benefits and a perception of certification being too easy or an unnecessary expense is enough to create scepticism.

Interestingly, a May 2023 social debate activity organised by an industry association representing corporate tenants and industry professionals in Hong Kong found by majority vote for a motion that certification was greenwashing and delivered no material benefits towards net zero-carbon by 2030. If that's what a snapshot of industry professionals are thinking, there's still a lot of work to be done.

ESG Investing as a Driver of Green Building Certification

A main driver of green-certified buildings is Environmental, Social and Governance (ESG) criteria. When investors are considering whether to put money into a company or its stock, they may look carefully at its ESG performance as another lens through which to consider non-financial performance of a business. A good rating will make the stock more valuable on the exchanges. Take Hong Kong as an example; its office market is increasingly competitive. How do you make your building stand out?

It must offer both operational efficiency and meet key ESG criteria.

Sustainability certification such as BEAM or LEED helps drive lower energy consumption; this is a competitive edge that can be shared with the tenant. The tenants may have set a net zero target as part of their ESG goals; this means their office energy use is included in their emissions measurement. Relocating to a lower energy building helps them achieve their ESG targets and in particular Net Zero which is often a 2025 or 2030 target date meaning there is only one potential office relocation before the target date.

From an investor's perspective, it also offers the prospect of future saleability and enhanced capital value in a business environment where sustainability is of increasing importance and where ESG reporting, including the need to show continuous improvement, is mandatory. This combination of increasing demand for green buildings, rent and price premiums and the downside risk of owning sub ESG standard buildings presents an investment opportunity. In the US, the National Association of Real Estate Investment Trusts (NAREIT) reported a 23% increase in occupancy, and 8% rental premium among green buildings. While in Canada the Government is backing schemes to retrofit buildings that can demonstrate a 50% reduction in emissions through its CA$200 million Deep Retrofit Accelerator Initiative (Day, 2022). Operational efficiency and lower emissions are benefits that are enjoyed by tenants of green buildings.

For developers and occupiers alike, it is not sufficient simply to claim a green building certification and hope that it will do. While there is substantial overlap, each standard is different and the purpose and emphasis of different standards needs to be considered alongside the local context, different stakeholders needs and desired outcomes.

Recent Trends: Regulations

In some respects the race to be green can be seen as a new element of city competitiveness. In Hong Kong, the Green Building Council defines a green building as one that "reduces the environmental impact of buildings and enhancing the health and well-being of building occupants by planning a focus on the planning and design, efficient use of resources (energy and water), waste and pollution reduction and enhancing indoor air quality" (Hong Kong Green Building Council, n.d.).

In green buildings, Hong Kong has been a relatively slow mover. As recently as 2019, the percentage of green building in Hong Kong was only 4%, compared to 30% in Singapore, 68% in London and 8% in Tokyo (Balaban & de Oliveira, 2019). This means that businesses with Net Zero targets might find more suitable sustainable office location options in Singapore than Hong Kong.

For Hong Kong, reducing building-related emissions is a critical part of meeting its own sustainability goals, specifically a 2050 net zero target. Hong Kong is one of the world's most densely populated cities and, according to the Electrical and Mechanical Services Department of Hong Kong, building-related activities accounted for 92% of electricity consumption in 2011, this compares with 40% in the EU (European Parliament, 2023). In Hong Kong, and presumably much of Asia, 50% is related to air conditioning in office and industrial buildings according to research led by Ir Professor Wang Shengwei and published on The Hong Kong Polytechnic University website. Improving the environmental quality of buildings, and performance of air conditioning in particular, is critical to reach Net Zero.

A comprehensive report published by Hong Kong 2050 is Now, a platform that supports collaboration, action and education to address the climate crisis, and the Greater Bay Area Green Finance Alliance highlights a number of next steps for Hong Kong buildings to decarbonise which are summarised as follows:

1. A real estate specific decarbonisation roadmap with actions and timelines.
2. Regulations to include the energy code and adoption of new technology, including sub-metering and the banning of low-efficiency machinery.
3. Market incentives to include post occupancy energy performance validation.
4. A comprehensive framework for retrofitting existing buildings.
5. Emissions trading schemes.
6. Increase renewable energy generation capacity.
7. Carbon tax.
8. A cross-agency body to co-ordinate relevant government departments (Lu, 2020).

According to a survey "On Green Building in Hong Kong" by Ophelia Chu ka-wai and Richard Cheung Yun-hing in 2017, there were a total of 53,432 buildings in Hong Kong, of which 25% were publicly owned and 75% privately owned. Of these, 1.5% were BEAM Plus certified and 0.34% LEED-certified. By early 2023, 1600 buildings had attained BEAM + certification (Ng, 2023), accounting for 3%. This represents a notable improvement compared to 2017 but still a small proportion of the total. Developers in Hong Kong may be reluctant to certify their existing building portfolios due to cost considerations, lack of government incentives, limited market drivers and having already met building codes and other regulations they may feel they have met their societal obligations. With an uncharacteristic period of oversupply looming from 2024, things may change and being green may lend a competitive edge.

Globally, according to the World Economic Forum, Asia accounts for 50% of total emissions, the need for leading economic cities in the region to take the lead in decarbonising the built environment is not only clear but an imperative to meet the targets that they themselves have voluntarily set.

Recent Trends: Working from Home

The COVID-19 pandemic has greatly changed ways of working across the world. This will profoundly affect the building industry. Working at home part- or full-time fits many aspects of ESG. Such hybrid work supports work-life balance and can result in lower stress and reduced anxiety. Businesses and employees have learned how to function while working from home and invested in the supporting technology.

The key future hybrid "resilient working space" and "home office" arrangements may include the following:

- Work that must be on site and for key workers, some flexible time options with resilient workspaces.
- Work that is free from location constraints may switch to remote virtual offices or combine a few days of office time and work from home for the remaining time.
- The pandemic also revealed drawbacks of traditional office designs, such as high density of common areas and poor ventilation. These often relied on short-term fixes, like regulating behaviour and intelligent cleaning and property management than longer term solutions like different office layouts with more space for employees.

Other post-pandemic office designs will include a much greater emphasis on indoor air quality—something that office buildings in mainland China have adopted for some time due to city smog: one-way office circulation, no-touch (or low-touch) facilities and doorways, "sneeze shields" in open plan offices and intelligent cleaning. Meeting expectations of employees will become an important part of attracting and retaining talent.

By extension, green criteria should be applied to the homes/apartments when people work from home. Such home-related emissions are part of scope 3 (supply chain) emissions alongside commuting and business travel, among others.

Overview of Green Building Certifications

Almost every country has their own certification system. Globally, there are more than 600 different green buildings certifications, including nearly 100 in the U.S. alone (Doan et al, 2017). They are awarded by

Green Building Councils (GBC) that have been set up in different countries. It is the industry's most direct response to the environmental crisis but interpretations may vary.

There are consistent characteristics. All green building certifications cover indoor environment: energy: building materials: and neighbourhood development. In addition to the three mentioned above, the principal ones are LEED Canada: HQE—High Environmental Quality, in France: Australia Green Star: Green Star NZ, of New Zealand: CASBEE in Japan—comprehensive assessment system for building environmental efficiency: and Green Mark of Singapore.

If we look around the world, we can see that, while broadly the certifications use similar criteria, they differ according to local needs and environments.

Certification methodology varies between new and existing buildings as well as between certification systems. This makes direct comparison very difficult. Some systems take a score-based approach that gives points for each environmental aspect*, or category, of the scoring criteria, with points awarded for meeting each aspect requirements. In contrast, in a performance-based system, the rating depends on the actual performance of the building in key aspects, including energy, water efficiency and materials, compared to a baseline of either a comparable existing building or energy standards applicable for the location (Civic Exchange/HKGFA).

* In the context of sustainability management activities, products and services of an organisation that interact with the environment are referred to as "aspects". An aspect may have positive or negative environmental impacts.

In addition, many certifications are voluntary while others are mandatory. For example, in California, a state short of water, the certification gives a higher weighting to water conservation. The California Green Building Standards Code (CALGreen), was the first state-wide green building code in the US and is a mandatory requirement for State organisations. In Japan and New Zealand, some local governments require certification by CASBEE and Green Star NZ respectively.

Here is the view of Melanie Alshab on certifications. She was COO and CDO of NWP Property:

> For the Indonesia ESG certificates, it has been a haul. To say that ESG certifications have a long way to go to be inclusive in emerging markets is not an understatement. Everything from the size of the asset to assessment

> criteria is unbalanced. Regardless, I personally support ESG and we had the IoT data collection and monitoring in place for management oversight for our geographically diverse portfolio. We made it a priority. However, almost every ESG investment we made had to have a financial return. We started working on this years ago before the broader private equity market woke up to ESG. So investments were made because they saved energy or because, in any market, it is bad to poison people with bad water. It did not cost us much to get the certifications. We already had all the data monitoring in place (IOT and other sensors all reporting back to our data repository) and water/energy conservation and social programs already in place.

Alshab added: "We ended up using GRESB and EDGE. Initially I was not a big fan of Edge but it was the only certification where we could do every asset in our portfolio, which includes small, medium and large retail malls and logistics assets – first, middle and last mile. Crucially, Edge provided a framework that could be applied to every assets and it is cost effective. In 2025, 8 years after our first Edge certification, we were able to secure a green loan from IFC, the first of its type in Indonesia. We used GRESB as we deal with Singapore banks almost exclusively and this helps us with a different type of financing".

Another local adaptation of standards is in earthquake-prone Japan, where the certification includes resilience to withstand earthquakes. Italy is a treasure house of ancient buildings—so GBC Italia has a certification for them (GBC Italia, n.d.). Lastly Green Star Australia only gives a six-star rating to fossil fuel free buildings (GBC Australia, 2020).

Table 9.1 gives a summary of more examples:

Of these, EDGE, WELL, RESET and NABERS are different from the others. "EDGE is a green building certification system focused on making buildings more resource-efficient" (GBCI, n.d.)—specifically embodied energy in materials, water efficiency and reduced energy use. It requires a minimum of 20% reduction in resource use compared to local market benchmark and is administered by Green Business Certification Inc. for the International Finance Corporation, a member of the World Bank.

WELL is a people-centric certification that looks at the health and safety of workplaces considering a range of metrics. The core concepts include air, water, light, nourishment, thermal comfort, movement, materials, sound, community and mind. This occupier focus makes WELL stand out, but it also means a certain immediate loss of business advantage among building owners. For the occupier, WELL certification may be

Table 9.1 Selected green building certifications

Name	*Jurisdiction*	*Rating*	*Year of Establishment*
Green Star	Australia	1 to 6 stars	2003
RESET	Canada	Modular criteria with audited performance data	2013
Green Building Evaluation Label	China	1 to 3 stars	2008
EDGE	IFC	Certified, Advanced, Zero Carbon	2014
BEAM and BEAM Plus	Hong Kong	Bronze, Silver, Gold, Platinum	1996/2010
IGBC	India	Certified, Silver, Gold, Platinum	2001
Greenship	Indonesia	Bronze, Silver, Gold, Platinum	2010
CASBEE	Japan	1 to 5 stars	2001
GBI	Malaysia	Certified, Silver, Gold, Platinum	2009
SEED	Pakistan	Silver, Gold, Platinum, Titanium	2016
BERDE	Philippines	1 to 5 stars	2009
Green Mark	Singapore	Certified, Gold, Gold Plus, Platinum	2005
G-SEED	South Korea	Green Grade 1 to 4	2002
EEWH	Taiwan	Qualified, Bronze, Silver, Gold, Diamond	1999
TREES	Thailand	Certified, Silver, Gold, Platinum	2010
BREEAM	UK	Pass, Good, Very Good, Excellent, Outstanding	1990
LEED	US	Certified, Silver, Gold, Platinum	1998
WELL	US	Bronze, Silver, Gold Platinum	2014
LOTUS	Vietnam	Certified, Silver, Gold, Platinum	2010
NABERS	Australia	1 to 6 stars	1998

Adapted from Savills Research and Consultancy, 2022

part of a staff engagement strategy to demonstrate well-being and provide assurance to staff that the new office smell is not derived from harmful chemicals. This is a common and understandable concern in some locations. The result is employee satisfaction and talent retention—making this a "nice to have" rather than a must have.

NABERS is short for the National Australian Built Environment Rating System and aims to provide "simple, reliable and comparable sustainability measurement" focusing on environmental performance (Federal, State and Territory Governments of Australia, n.d.). NABERS measures actual performance to evaluate building sustainable performance based on water, energy, waste, lighting and air quality. It is a performance-based rating system that uses data collected over a 12 month period (Walsh, 2023). To achieve NABERS 6 star rating, buildings have to be fossil fuel free, this includes restaurants and food courts which must adapt to full electrification. This can be challenging for grilling, flash frying and baking.

According to RESET their range of standards and tools focus more on data to rate sustainable and healthier built environments. By using data quality and ongoing monitoring the result is continuous improvement through data-driven management. RESET offers different standards that can be suitable for different approaches (RESET, n.d.). The full range of RESET modular standards cover embodied (carbon, circularity and health) and operational standards (air, energy, water and waste). This modular approach to certification is likely to gain more traction, allowing more specific focus on what matters most to each particular organisation.

There is also Global Real Estate Sustainability Benchmark (GRESB) which is used by asset managers and property developers as a portfolio assessment benchmark. It considers a range of both qualitative and quantitative data to score and compare a portfolio of assets against their peers. The annual assessment includes options for existing and development assets with a management component to assess the management tools being used.

Now let us look in more detail at some of the leading certifications. Despite some consistencies, direct comparison is difficult because of differing criteria, weighting and methodology. The following summarises some of the key aspects.

One: Leadership in Energy and Environmental Design (LEED, U.S.)

One of the most important certifications in the world is LEED, established in the United States in 1998. It is the closest to a global standard, since it has been adopted in 160 countries.

LEED gives energy use a 35% and 30% weighting for existing and new buildings respectively. That compares with 24% and 29% respectively for Building Environment Assessment Method (BEAM), a system created in Hong Kong in 1996. LEED gives credit for use of off-site renewable energy and carbon mitigation. LEED makes disclosure of energy data to the certification body a must; BEAM Plus give credits for disclosure and sharing with building occupants but does not make it a requirement. LEED pays attention to water efficiency (Chu & Cheung, 2017).

In sourcing and manufacture of materials used, LEED allows a maximum of 160 km from the site location; BEAM Plus allows 800 km. LEED specifies certification of materials and waste management. BEAM plus focuses more on materials used, such as chlorofluorocarbons (CFCs) in cooling systems, use of timber in temporary structures during construction and efficient use of materials. LEED offers 13 points for materials and BEAM Plus 22 and a bonus point. In indoor air quality, LEED offers less than half the number of points than BEAM plus. It also considers tobacco smoke; BEAM Plus focuses on control of Legionnaires' Disease.

One builder who has chosen LEED is Henry Cheng, co-founder and Chief Executive Officer of the Chongbang Group. Chongbang is a real estate investment and development group of companies, founded in 2003 by a consortium of Hong Kong and Singaporean investors and a group of seasoned real estate professionals. The group has developed a portfolio of urban mixed use Life Hub projects in mainland China. According to published interviews, Cheng chose LEED certification because it is internationally recognised and offers the opportunity for continuous improvement. So far, six of its projects have achieved LEED Certifications.

This is what Cheng said in an interview in October 2021published on LinkedIn: "LEED, as the most widely recognized rating system for green buildings and cities in the world, provides a set of performance standards against which the sustainable operation of our projects can be measured. The LEED framework has helped us identify opportunities for continuous optimization across project procurement, energy consumption, water use and waste management, as we strive to achieve sustainable and green operations and maintenance through continuous operational enhancements, energy audits and green building education, creating a better living, working and shopping environment for all".

> First of all, energy consumption is a major issue. The HVAC system in our project has been responsible for a significant proportion of total energy consumption. Building upon the LEED strategies, we strive to ensure optimised energy efficiency and cost savings through retro-commissioning of existing buildings, such as retro-commissioning both air-side and water-side systems in HVAC and installation of an intelligent energy management system. Secondly, in order to create a healthy and comfortable space for customers, achieving energy savings without compromising their experience, we continue to seek improvements through tracking and monitoring the indoor air quality, humidity, light, temperature and sound as well as conducting follow-up satisfaction surveys on tenants and customers.
>
> Engaging tenants in going green is another major challenge. We included 'green leasing' and 'green fit-out' clauses in tenancy agreement so we can work with tenants and reach the goal of reducing energy consumption through cooperative efforts. Secondly, we place great importance on site selection. Take one Life Hub for example. The location is served by a subway line and eight regular bus routes, and is in close proximity to the railway station, which is easily accessible by public transportation. Chongbang has also put efforts into creating a sustainable and people-friendly environment to make it easier for the projects' retail tenants, office tenants and shopping customers to walk or bike, such as planning separate lanes and designated spaces, installing convenient facilities and benches in pedestrian zones, in order to further enhance the living, shopping, working and entertainment experience for this community.
>
> The other two areas are related to energy efficiency, specifically water efficiency management and waste management. The two Life Hub projects mentioned above both have had water-saving sanitary appliances installed. Additionally, we have developed and followed the "collection - sorting - transport - processing - reporting" procedure, with clear accountability. Chongbang's projects will start to adopt the ISO 14001 environmental management standard and continue to step up employee training to improve waste utilisation and recycling.
>
> The incremental cost has been added to the overall cost of these projects. But look at it from the perspective of project life cycle, they are just a small portion. To continue operating in a sustainable way that aligns with LEED standards and requirements, we will adopt measures across our projects such as setting energy efficiency, conducting periodic energy audits and retro-commissioning, as well as installing an energy management system

etc. to achieve energy optimization. Adopting these measures may incur costs in the initial phase, but will also yield considerable environmental benefits in the mid-term and long term.

May Lv agrees, saying that green certification helps deliver higher efficiency over the whole life cycle, energy, water, material conservation. Examples she gave include 30% water savings, reduced maintenance costs of 25–30% by using renewable materials, overall operating cost savings of 30% and value premiums of 10–15%.

Put simply, green certification supports better building management and drives financial performance. This chimes with a point made by Andy To, "green certification is all commercially driven, there are some government concessions and incentives, but tenants expect it and will not consider buildings that are not green certified. Some Chinese owned retail brands are now using LEED certification as standard, including Luckin' Coffee and brands under the Anta Group". He added that some markets like Japan are a bit slower, but Japanese have a tradition of sustainable design, such as using timber for secondary structural elements and Japanese developers tend to be rather humble in broadcasting their achievements. While in South Korea, To points to the international funds active there as a driver for LEED certification including offices, warehousing and retail property".

Two: Building Research Establishment Environmental Assessment Method (BREEAM)

This was established in the U.K. in 1990 and is the earliest established green building certification system. It is owned by BRE, a profit-for-purpose organisation whose purpose is to "contribute to a thriving and sustainable world by developing science-led solutions to built environment challenges" (BRE Group, n.d.).

In 2017, BREEAM had 7 times more certified buildings, in 77 countries, than LEED, in 160 countries. So BREEAM has been used to certify more buildings but LEED has been adopted globally (Doan et al, 2017). Like LEED, BREEAM focuses more on environmental factors, including site, water, materials and IEQ, and the life cycle of a building.

LEED uses more assessment categories, measures improvement and adds up the total to score; this means that it is considered more transparent. BREEAM and Green Star NZ, a certification used in New

Zealand, set absolute criteria and use pre-weighted scoring categories. Green Star NZ uses different weightings depending on the type of building—industrial, office and educational. Assessing green criteria may use international standards or local equivalent standards. LEED tends to require international standards; BREEAM allows local equivalents. At the same time, BREEAM is stricter in setting absolute requirements to score points.

Three: Building Environment Assessment Method (BEAM)

This certification was drawn up in 1996 in Hong Kong, with Beam Plus added in 2010. It is one of the Hong Kong Green Building Councils (HKGBC) means of assessing sustainability performance of buildings. BEAM + has been used outside Hong Kong in Macau, Shenzhen, Guangzhou, Shanghai and Beijing. It has seven criteria:

1. Integrated design and construction management
2. Health and well-being
3. Sustainable site
4. Materials and waste
5. Energy use
6. Water use
7. Innovations and additions

The key differences between BEAM Plus and LEED are in water, materials and indoor air quality. While BEAM Plus uses multiple targets around potable water quality and treatment of wastewater, LEED pays attention to water efficiency (Chu & Cheung, 2017). One example is that BEAM pays attention to monitoring water leaks while LEED does not. Another example is the LEED toilet flush standard compared to BEAM Plus which makes allowance for needs based differences in water used in male and female toilets.

The two certification standards also differ when it comes to considering building materials. One notable difference is sourcing and manufacture of materials used; LEED allows up to 160 km, BEAM Plus allows 800 km from the site location. Another difference in this aspect is that LEED specifies certification of materials and waste management while BEAM Plus focuses more on materials used e.g. CFCs in cooling systems, use of timber in temporary structures during construction as well as efficient use of materials. Overall, LEED offers 13 points for materials and BEAM Plus 22 and a bonus point.

The final major area of difference is indoor air quality (IAQ). BEAM Plus has more than double the number of points for this aspect than LEED; but there are different credit categories in this section. One notable difference is that LEED considers tobacco smoke while BEAM plus focuses on control of Legionnaires' Disease.

From a building occupier standpoint, LEED offers the closest thing to an international standard that corporates can refer to across markets and avoid difficult comparisons. A global standard may specify "LEED or local equivalent" as the requirement for Group approval for a relocation. This then dovetails with an office or retail green fit out, or interior renovation, standard which depends in key areas on the overall building certification.

In contrast to the previous three, the Japanese CASBEE standard takes a broader contextual approach from design stage to later renovation. It also offers CASBEE Urban Development for cities. It considers internal and external rather than "green categories". This may be considered as built environment quality (Q) and built environment load (L). CASBEE is the only rating system with an earthquake resistance sub-category: it was also the first to offer a mixed use development area assessment tool.

The Way Forward

It is not expected that a single certification system will dominate but that there will be standardisation of measurement of a series of common metrics across a broad spectrum of material factors which can then be isolated and presented separately. This will be in addition to the local star or medal rating that covers so many factors.

For those aiming for zero-carbon emissions, energy efficiency is the single most important objective; the Net Zero imperative means switching to renewable energy tariffs is an overarching goal.

One basic benchmark to consider when making comparisons between buildings is energy intensity—the amount of energy used per square metre or number of occupants; for that, data transparency is essential. Improving publicly available energy data for individual buildings is a key step. Maximising energy efficiency also requires collaboration with tenants and this is an area where rapid progress can be expected over the next 3–5 years.

It is interesting to note that some retail tenants are already developing their own interior standards, including Kering, who have an extensive guide for store renovation and operation standards that go beyond any current certification and are tailored for the luxury retail sector. The LVMH "Life in Architecture" guide is an open source document "aiming to unite the entire industry, including competitors, around the challenges of eco-design for buildings and boutiques" (LVMH, 2025) (Table 9.2).

The above list reveals the biggest drawback of Green Building Certifications; they are designed around common factors that may not apply to everyone. As the Kering case reveals, when you are a retailer, you are

Table 9.2 In-house green store standards for selected retail brands

Retail brand	*Certification name*	*Focus areas*	*Remarks*
Kering	Kering Standard for Stores	Store design, engineering, construction, materials, certifications and documentation	Approx. 50% of points are energy related
LVMH	LIFE in Architecture Awards	Building insulation, lighting density, air quality, energy, interior design, maintenance	Generally equivalent to or exceeds LEED
Hermes	Harmonie	> 80 criteria related to carbon footprint, air quality, local sourcing, biodiversity and environmental health	Initial assessment of sustainability potential of base building and actual performance after completion
Starbucks	Greener Store	25 standards covering water stewardship, partner engagement, energy efficiency, waste diversion, renewable energy, responsible materials and sites and communities	Developed with external partners
Estee Lauder Companies	Green Building Standards & Responsible Store Design	Energy, water, waste, and indoor air quality. Additional initiatives by Retail Sustainability Leadership Council, covering: material selection, furniture, operations and packaging	Features energy management system and sustainability champions to support efforts

(Kering, 2019; LVMH, n.d.; Hermes, n.d.; Starbucks, 2023; Estee Lauder Companies, 2023; Higgins, 2023)

already self-selecting an accessible location. You do not need a green certification to confirm that, there are other more important things to focus on. The Hermes Harmonie Standard is a dynamic one, it considers the maximum potential score of the individual site based on a review of the base building and then sets a maximum available score for that specific building. If you are renovating a historic landmark building, the sustainability potential is very different to that of a modern, newly finished shopping centre. This is where the RESET modular standards could be particularly helpful in supporting an adaptive approach to impactful sustainability targets.

Green Certification Successes

Andy To highlights the success of green certification in Beijing CBD which he says it he mostly densely populated LEED district in the world. Planning for the area started in 1993 and obtained State Council approval. It is now home to over 4000 foreign businesses, more than 29 projects are LEED gold or above and accounts for 25% of all the green-certified buildings in Beijing and the whole of the core area is LEED Gold Neighbourhood Development (Chaoyang District People's Government, n.d.).

Of interest to some readers will be the existence of United States Forces Korea (USFK) Regulation 201–1 dated 30th July 2020. This 269 page publicly available Environmental Governing Standards regulation "provides specific criteria and management practices for environmental protection on USFK installations". Contents include air emissions, drinking water, waste water, hazardous materials and waste, medical waste, petroleum, oil and lubricants, noise, pesticides, historic and cultural relics, natural resources and endangered species and more. It is the most comprehensive set of sustainability standards found in researching this chapter. It does not have a specific building standard but the 2024 opening of new housing towers at Camp Humphreys noted the LEED gold certification (Napolitan, 2024).

LEED silver is a minimum requirement for US Army Corps of Engineers Far East District (USACE FED) according to a quote in the same article. A USACE document dated September 2014 offers comprehensive guidance for LEED v4 implementation and applies to all US Army facilities worldwide (new buildings and renovations only) (USACE, 2014).

It is worthy of note that as early as 2001 the US Army was using the Sustainable Project Rating Tool (SPiRiT).

Summary

Like all sustainability standards, Green Building Certifications are still evolving. Currently, there are many different standards, measurements and weightings. Which one should a building owner or tenant choose, one or two? The advice of professionals is to "make the decision (on which certification) at the outset of the design process rather than midway through the construction".

Different jurisdictions have different emphases. But does earthquake resilience really belong in a green building certification or should there simply be in a distinct earthquake resilience certification? Probably the latter. How should a building owner be given credit for maintaining an older building and making it more sustainable rather than redeveloping the site and getting a maximum rating with a new one? Certifications do not give credit for retrofitting rather than redeveloping but the RESET modular approach does enable targeted performance improvement in specific areas.

In the rush to meet green certification standards, are building owners missing out on financial savings? According to Lizette McNeill at IXO Partners "companies are making changes to meet certification requirements, not realising that there are so many other opportunities for real cost savings from energy and materials".

Green building certifications are the first step to take. Used effectively over a period of time, they are definitely a valuable tool to help building owners and users improve sustainability performance. In this sense they are an essential step and a means of gaining recognition for their efforts, but are they too easy? Those retailers that have come up with their own green standards seem to think they can be improved on, notably adjusting the standards according to the nature of the base building and ignoring some of the factors altogether. Ultimately, when it comes to climate change, energy efficiency and the net zero pathway is the imperative and this could and should be given a separate recognition. Is there any reason why building energy performance data shouldn't be published so tenants and stakeholders can make their own judgement? This is where the operation certifications are important and this is why the certifications themselves are constantly evolving. If green certifications enable buildings to wear a green cloak to hide poor energy performance they are of no benefit to people, planet or profit.

References

Balaban, O., & de Oliveira, J. A. (2019, June). *Hong Kong Engineer*. Hong Kong Engineer. Retrieved August 29, 2023, from http://www.hkengineer.org.hk/issue/vol47-jun2019/cover_story/

BRE Group. (n.d.). *About us*. BRE Group. Retrieved August 29, 2023, from https://bregroup.com/about-us-2/

The Business Times. (2023, March 27). Lloyds races to sell Chinese tycoon's Canary Wharf office loan. *The Business Times*. https://www.businesstimes.com.sg/property/lloyds-races-sell-chinese-tycoons-canary-wharf-office-loan

Chan, I., & Leung, M. (2017, December 18). *Sustainable development worldwide: Costs of green buildings* [Report for HKIS-funded Research Project]. Retrieved May 30, 2023, from https://www.hkis.org.hk/ufiles/HKIS17_GB_Report_20180301_R.PDF

Chaoyang District People's Government. (n.d.). *Beijing Central Business District*. Retrieved July, 2025, from https://english.bjchy.gov.cn/central-business-district

Chu, O. K., & Cheung, R. Y. (2017). *A comparison on two certification systems: Leadership in energy and environmental design and building environmental assessment method on green building in Hong Kong*. World Sustainable Built Environment Conference 2017 Hong Kong. Retrieved 2023, from https://www.irbnet.de/daten/iconda/CIB_DC30181.pdf

CIM. (n.d.). *Optimisation tips to improve your building's sustainability performance* [eBook]. eBooks & Guides. https://www.cim.io/documents

Day, K. (2022, July 6). *Why commercial property is more valuable when it's green*. MODUS | RICS. Retrieved August 29, 2023, from https://ww3.rics.org/uk/en/modus/built-environment/commercial-real-estate/why-commercial-property-is-more-valuable-when-it-s-green.html

Doan, D. T., Ghaffarianhoseini, A., Naismith, N., Zhang, T., & Tookey, J. (2017). A critical comparison of green building rating systems. *Building and Environment, 123*, 243–260.

Estee Lauder Companies. (2023, June 29). *ELC launches responsible store design program – The Estée Lauder Companies Inc.* Estee Lauder Companies. Retrieved October 16, 2023, from https://www.elcompanies.com/en/news-and-media/newsroom/company-features/2023/responsible-store-design-launch

European Parliament. (2023, March 14). *MEPs back plans for a climate neutral building sector by 2050 | News*. European Parliament. Retrieved August 29,

2023, from https://www.europarl.europa.eu/news/en/press-room/20230310IPR77228/meps-back-plans-for-a-climate-neutral-building-sector-by-2050

Federal, State and Territory Governments of Australia. (n.d.). *What is NABERS?* NABERS. Retrieved May 19, 2023, from https://www.nabers.gov.au/about/what-nabers

GBCI. (n.d.). *EDGE | Green building certification*. Retrieved May 18, 2023, from https://edge.gbci.org/

Green Building Initiative. (n.d.). *Why green globes? – Green building initiative*. Green Building Initiative. Retrieved May 19, 2023, from https://thegbi.org/green-globes-certification/

Green Building Council Australia. (2020, October 29). *News News What does a new 4, 5 and 6 Star Green Star building look like?* Green Building Council of Australia. Retrieved May 19, 2023, from https://new.gbca.org.au/news/gbca-news/what-does-new-4-5-and-6-star-green-star-building-look/

c. (n.d.). *GBC HB - Documenti*. GBC Italia. Retrieved May 19, 2023, from https://gbcitalia.org/certificazione/gbc-historic-building/gbc-historic-building-documenti/

Hermes. (n.d.). *A path for a high social and environmental performance in real estate*. Hermès Finance. Retrieved October 16, 2023, from https://assets-finance.hermes.com/s3fs-public/node/pdf_file/2023-08/1692976774/guide-referentiel_harmonie_en.pdf

Higgins, M. (2023, July 5). *Estée Lauder launches responsible store design program*. Environment+Energy Leader. Retrieved October 16, 2023, from https://www.environmentenergyleader.com/2023/07/elc-launches-responsible-store-design-program-for-sustainable-retail/

Hong Kong Green Building Council. (n.d.). *What is green building?* HKGBC. Retrieved May 19, 2023, from https://www.hkgbc.org.hk/eng/about-us/what-is-green-building/index.jsp

The Hong Kong Polytechnic University. (2020). *PolyU solution reduces energy consumption in buildings by up to 40% | Autumn 2020 - Issue 2 | Excel x Impact*. PolyU. Retrieved August 29, 2023, from https://www.polyu.edu.hk/publications/excelximpact/issue/202002/knowledge-transfer-entrepreneurship/polyu-solution-reduces-energy-consumption-in-buildings-by-up-to-40

Kering. (2019, March 9). *Kering standard for stores store planning and construction*. Retrieved October 16, 2023, from https://www.kering.com/api/download-file/?path=Kering_Standard_for_Store_Planning_and_Construction_c50c3866a3.pdf

Lu, L. (2020, December). *Decarbonising Hong Kong buildings* [Policy recommendations and next steps]. https://www.hk2050isnow.org/resource/decarbonising-hong-kong-buildings/

LVMH. (n.d.). *LIFE in stores awards - Initiative LVMH*. LVMH. Retrieved October 16, 2023, from https://www.lvmh.com/group/lvmh-commitments/social-environmental-responsibility/life-in-stores-awards-initiative-lvmh/

LVMH. (2025, May). *Life in architecture*. Retrieved August 22, 2025, from https://www.lvmh.com/commitment-in-action/for-the-environment/life-in-stores-awards

Napolitan, R. (2024, September 13). *New housing towers open at Camp Humphreys, celebrating legacy and sustainability*. Army.mil. Retrieved July 2, 2025, from https://www.army.mil/article/279654/new_housing_towers_open_at_camp_humphreys_celebrating_legacy_and_sustainability

Net Zero Asset Managers. (n.d.). *The Net Zero Asset Managers initiative*. The Net Zero Asset Managers initiative – An international group of asset managers committed to supporting the goal of net zero greenhouse gas emissions. Retrieved June 6, 2023, from https://www.netzeroassetmanagers.org/

Ng, J. (2023, March 23). *Green buildings in Hong Kong: How effective are they?* Earth.Org. Retrieved August 29, 2023, from https://earth.org/green-buildings-in-hong-kong-how-effective-are-they/

Oliver, J. (2023, May 11). 'Stranded assets': investors reckon with obsolete offices. *Financial Times*. https://www.ft.com/content/b9fae82d-2612-4d24-971e-ef610d5c701a

RESET. (n.d.). *Standard | RESET®*. RESET® Standard. Retrieved July 19, 2023, from https://reset.build/standard

Savills. (2022, June 1). *Is there a green premium, or a brown discount? | Themes*. Savills. Retrieved May 31, 2023, from https://www.savills.com/prospects/themes-is-there-a-green-premium-or-a-brown-discount.html

Specification Online. (2023, May 24). *Growing demand for buildings designed with BREEAM certification*. Specification Online. Retrieved June 5, 2023, from https://specificationonline.co.uk/articles/2023-05-25/glenigan/growing-demand-for-buildings-designed-with-breeam-certification

Starbucks. (2023, April 17). *What's inside a Greener Store?* Starbucks Stories. Retrieved October 16, 2023, from https://stories.starbucks.com/stories/2023/whats-inside-a-greener-store/

USACE. (2014, September). *USACE Army LEED® v4 implementation guide*. Whole Building Design Guide. Retrieved July 2, 2025, from https://www.wbdg.org/FFC/ARMYCOE/SDP/USACE_ArmyLEEDImplementationGuide-v4.pdf

UNEPFI. (n.d.). *UN-convened Net-Zero asset owner alliance – United Nations Environment – Finance initiative*. UNEP FI. Retrieved June 6, 2023, from https://www.unepfi.org/net-zero-alliance/

United Nations Environment Programme. (2009). *Common carbon metric for measuring energy use & reporting greenhouse gas emissions from building operations*. Retrieved October 16, 2023, from https://wedocs.unep.org/20.500.11822/7922

Walsh, P. (2023, March 11). *NABERS UK: Everything you need to know*. CIM. Retrieved May 19, 2023, from https://www.cim.io/blog/nabers-uk-everything-you-need-to-know

CHAPTER 10

The Changing Landscape

"The future of cities lies in their ability to harness the power of technology to solve complex urban issues and create sustainable, liveable communities" Michael Bloomberg.

Sustainability Tension

Business objectives remain firmly profit related. At the same time sustainability goals are seeing increasing emphasis, a degree of tension is inevitable due to the competing demands of financial, environmental and social sustainability (Van der Byl & Slawinski, 2015). The result can be described as "a dynamic relationship that involve both competition and complementarity" (Epstein et al., 2015). As has been mentioned in passing in some of the previous chapters, different sustainability goals may be intertwined or have tensions between them as well as with the overarching profit motive of the business. The result is a set of new, complex and dynamic leadership and management challenges that need to permeate throughout the organisation and beyond.

For evidence of sustainability tensions, look no further than the household name organisations that have revised, rewritten, restated or relegated their corporate ESG targets in late 2024 and early 2025. It isn't difficult to imagine the tension around the boardroom table as discussions

S. Crispin, *Sustainability in the Built Environment*,
https://doi.org/10.1007/978-981-95-6549-8_10

unfolded and decisions were made. Stories from "those in the room when it happened" are beginning to emerge. It is not uncommon either for a sudden and unexplained restructuring of internal sustainability teams to take place. Are these changes made to advance sustainability performance, improve the reporting process, save costs or side-line inconvenient truths?

Writing in the Harvard Business Review, Kell et al., 2025 describe the tensions of politics, populism and polarisation against people, planet, profit. While some public unravelling of sustainability has taken place, they remain optimistic for the sustainability case due to decarbonisation of electricity. The IEA states that renewables will generate nearly half of global electricity needs by 2030 (IEA, 2024). At the time, the reality of increasing climate change-related disasters changing popular opinion and the business imperative for sustainable and profitable business models. One anonymous interviewee mentioned a circle of tensions, or misalignment of tensions, between fund manager, investors and asset managers, between investors and building owners and among investors from different geographies. Finding common ground before anything can be done is a challenge in itself. Despite all this the imperative remains. Businesses are continuing to pursue sustainability strategies by investing in technology to drive efficiency, decarbonise in the expectation of future regulatory changes and gain a competitive edge by being fit for the future (Kell et al., 2025).

This chapter turns to focus on the tensions resulting from adoption of sustainability objectives, challenges some of the accepted wisdom and the changes in organisational behaviour including data collection and the role of technology. This includes the tenant or occupier perspectives and the requirements for adaptation of existing buildings to meet tenant demands. One example mentioned in discussions with interviewees was the installation of EV charging stations resulting in increased overall energy consumption for the building.

A market flash from APREA in the second half of 2024 helps set the scene of activity:

1. ESG standards are transforming the market particularly for new developments.
2. New value propositions are being enabled by operating efficiencies resulting from smart building to PropTech.
3. ESG compliant properties are a key requirement for both occupiers and investors and this is driving retrofitting of existing buildings.

4. Demand for green logistics, industrial and data centre properties is resulting in increased demand for renewable energy.

One example of sustainability tension may be post-acquisition integration. A takeover by a multinational organisation of a small business may require the small business to adopt the sustainability standards of the multinational. This is a subject possibly overlooked at the pre-acquisition negotiation stage and it then becomes an unwelcome surprise to the new acquisition that the Grade B, cost effective but "brown" offices, do not meet the corporate standards. Upgrading to a green building almost certainly increases costs which will almost certainly impact the management team bonuses of the now new "division", or the additional costs have to be "grandfathered". Either way, such sustainability tension is likely to be unwelcome during the integration process.

Finding ways through the maze of sustainability tensions is one of the invisible arts of the sustainability leader and helps redefine organisational behaviour.

Organisational Behaviour

When companies set a sustainability strategy, they start a process of organisational change that they may not always fully comprehend at the outset and, as a result, may not be prepared to follow through on and fully resource. The resulting sustainability tension can lead to an element of dysfunction as the trajectory set at organisation headquarters cascades throughout the organisation. Regional operations may or may not have been consulted as part of the process and remote offices are left to work out for themselves how to implement them and how to allocate any additional costs.

Organisation sustainability strategies often evolve around the Sustainable Development Goals (SDGs). The selection of these begins with materiality assessment which is a process of asking internal and external stakeholders what they think are the relevant sustainability issues to the organisation. Stakeholders include staff, executives, shareholders, clients, suppliers, community representatives and so on. Double materiality considers both the internal and external impacts the organisation has on people and planet and the financial materiality that sustainability and climate has on the organisation. It is unlikely that all stakeholders will have the same expectations, tensions will be seen even at this early stage. The organisation will have to moderate the results of the materiality assessment and use a blended approach to develop the strategy.

Materiality assessment is followed by baseline assessment, establishing target setting principles, target identification, target setting and establishing a governance and oversight structure to oversee the implementation. For most organisations this is a new process that might involve more "navel gazing" introspection than they are used to or comfortable with. The initial SDG-linked objectives might be left vague at this stage, such as "work to minimise GHG emissions", and only later tightened up to be firm, measurable commitments, such as "achieve net zero across our global property investment portfolio by 2030".

The goals can add unwelcome costs and have a disruptive impact on the business including retrofitting offices to meet certification standards and improving energy efficiency. Implementing single use plastic free office policy, for example, can force unwelcome changes to staff as set out below:

1. Communication plan to let everyone know and understand.
2. Switch from bottled to tap water. Install a filter as required.
3. Reduce use of single use plastic stationary.
4. Reduce toilet plastic use, e.g. liquid soap.
5. Eliminate single use plastic cups and cutlery.
6. Switch to reuseable food containers, including for takeaway food.
7. Reuse waste as far as possible, e.g. plastic bags.
8. Recycle whatever cannot be reused (Bernoville, 2020).

Even diverting E waste from landfill by donating old IT equipment to local schools or charities incurs a cost for IT colleagues to spend time wipe hard disks one computer at a time to abide by data security policies. This kind of social action in donating time and governance action in data security might be an easy thing to skip when leadership sustainability tension emerges when preparing budgets for the year ahead.

In the case of property investors, Dhawal Doshi at NWP in Indonesia confirmed the complexity of changing organisational behaviour when implementing building sustainability. "Suppliers and operating teams have all had to adapt, this includes tenants who now have meters in places that they did not have them before. It takes work to explain the rationale, train people, talk to tenants, change the way we charge for AC while top management support is essential as well as giving people time to adapt to the new reality. Having achieved such a large green portfolio, our attention is now 80% monitoring and improvement of already green assets". Tenant engagement can be a time consuming business and some organisations only allow their leasing teams to have direct communication with tenants. This means that the sustainability teams have to first educate their client facing colleagues before tenant engagement can even begin.

Organisational Governance and Decision-Making

The COVID-19 pandemic was a dual financial and public health crisis that resulted in a growing focus on corporate sustainability. As we saw in Chapter 4, the built environment, public health, climate change and sustainability are interlinked resulting in even greater pressure on businesses in the built environment value chain. Investors are concerned with resilience and long-term value creation making sustainability part of good corporate governance, risk management, transparency and accountability frameworks and increasingly an essential part of boardroom decision-making.

Key sustainability factors board directors need to pay attention to:

1. Regulation.
2. Litigation risk.
3. Reputation.
4. Stakeholders.
5. Reporting requirements (WBCSD, 2020).

Directors responsibilities include corporate governance "a framework by which various stakeholder interests are balanced and efficiently and professionally managed" (Committee on the Financial Aspects of Corporate Governance, 1992). This includes trust, good faith, confidence, loyalty and scrupulousness.

In some industries, a failure to take into account ESG and broader sustainability-related risks, could amount to a breach of fiduciary duty, even in markets, such as the United States, where director responsibility is solely or mainly to the shareholder. Increasing attention is being given to the interests of broader stakeholder groups, not just focused on shareholders. This includes anyone impacted by the business operations including customers, suppliers, employees, the community and the environment. Acting in the best interests of the company means taking account of stakeholder interests in the long as well as short term. This means reputation and standards of business conduct which fall under the ESG reporting framework.

Directors are assumed, and expected, to have the skill and knowledge to fulfil their role and discharge their duties. They should have access to information and resources when needed. This includes ESG matters in decision-making. While this may mainly involve non-financial reporting, climate change risk is considered financial risk. With increasing regulatory requirements around decarbonisation and sustainability reporting, the role of the board in managing expectations is only going to increase (WBCSD, 2020).

ESG Data

While the general practices around ESG may have already been in place in different forms at many organisations for years, reducing environmental impact, corporate social responsibility, employee training and compliance with tax regulations, data privacy and so on, the systematic ESG reporting approach and data collection is relatively new for most organisations. Even for those with some years of doing so are likely to find the ever increasing needs for better quality data and more extensive data, for example, scope 3 reporting requirements, increasingly burdensome. At times of economic slowdown of political pushback, it's easy to just revise targets than stick to be plan.

The following is from a Deloitte survey on scope 3 data collection:

1. Supplier engagement: It is critical to involve suppliers in emissions reduction efforts.
2. Data accuracy: Reliable data collection and reporting are basic requirements for effective emissions management.
3. Data reliability: Obtaining data from suppliers is a significant hurdle.
4. Engagement difficulties: Securing suppliers support for shared emissions reduction initiatives is a challenge.
5. Practical tools: Strong demand exists for resources to support secure and accurate data collection.

6. Sustainable resources: Securing renewable electricity tariffs and low carbon materials is critical.
7. Regulations: regulatory clarity is not an immediate concern (Hansor, 2025).

According to Robert Hansor, Director, Sustainability at Deloitte, "Generating accurate data insights is critical otherwise companies risk building a strategy on shaky ground, and renewable electricity procurement is a key accelerator for Scope 3 progress".

Such data is not widely available. Raefer Wallis at RESET comments on his work in retail interiors, "only 0–1% of materials by weight come with product specific information in China. Europe does slightly better but not by much. Office materials are more like 0–3%. This is a huge issue and we've had to rely on proxy data which are on the conservative side and are a means to encourage suppliers to provide the data so they are not penalised". This is how scope 3 reporting works, if more suppliers take up the idea of product passports they should win more sales.

ESG reporting data needs to be collected from different entities within the organisation, no part of the operation is excluded. Data needs to be evidence based to be of auditable quality. The data needs to be held securely requiring investment in technology, it needs employees with the expertise to analyse it in order to fully understand, communicate and benefit from the findings. There may be a need to adapt operations and stay abreast of regulations across multiple jurisdictions, the scale and scope of which can seemingly change with the wind.

According to KPMG, 75% of organisations must now, or will soon have to, report ESG data. Of those, only 29% have the resources in place to meet ESG assurance requirements (KPMG Australia, 2024). With most companies self-reporting, common data difficulties include:

1. Multiple sites in many different locations.
2. Overlapping data points.
3. Multiple data input staff.
4. Low levels of checking.
5. Random audits only.
6. Qualitative nature of social sustainability metrics.

For the built environment, one common problem is the lack of a common definition of a square metre which is a problem comparing metrics in different jurisdictions. For example a gross square metre may or may not include the external wall, it may include balconies or part of balconies, stairwells too may be included. A contract for an office in Hong Kong includes a share of the common areas while in Japan this is not included, for a 10,000 sq.m. office the result may be 1,000 sq.m. more of useable floor area in Japan compared to Hong Kong.

Similarly the carbon intensity of grid electricity (measured in grams of carbon dioxide equivalent per kilowatt-hour), or the emissions factor, can vary widely by country and by the time of day. Carbon intensity of electricity in Japan was over 582 g CO2/kWh in 2023 compared to 475.45 in Vietnam. It gets even more complex because emission factors for renewable energy vary during the day, depending on sunshine and wind variations.

Data security is a governance issue and ESG data can be sensitive in some situations. Investment banks may be reluctant to share electricity data with their landlords, for example, because it may be taken as an indicator of how busy they are, perhaps working on a sensitive transaction or major IPO. A resource intense retail shop may not want to share the data with the landlord for fear of not being offered a lease extension in favour of a "greener" tenant. Anecdotal evidence suggests that even retail tenants with green stores are reluctant to share their data with mall owners. Employee data is always sensitive. Any data breaches can impact organisation reputation.

As the examples above show, when it comes to scope 3 supply chain data, things get very complex. Requesting an upstream or downstream business partner to share their data is not a simple thing. Security for supplier shared data is even more important. Governance practices for ESG data include embedding data privacy into the governance framework, accountability for data security, transparent risk mitigation to build trust with partners and vendor management for IT suppliers.

The Role of PropTech and AI

Leaving a self-learning artificial intelligence building management system in full charge of improving sustainability performance could result in the doors being locked and energy using systems being switched into hibernation mode. This section aims to explore the opportunities that technology and AI have in supporting decarbonisation of the built environment. The combination of property and technology is frequently combined into the term "PropTech". According to a Savills report, a PropTech strategy comprises Internet of Things (IoT) devices such as sensors and monitors and data management software to collect and manage the data from the IoT devices (Sellers, 2022).

A building technology strategy should be able to reduce energy consumption, monitor water, waste and air quality, optimise resource management, support urban design, improve transparency and analyse data for preventative maintenance (Proptechbuzz, 2025). Yet according to Jonathan Hannam, Co-Founder and Managing Director of Taronga Group, "real assets, the world's biggest asset class, is also one of the slowest at adopting new technology". Taronga has a business model that supports technology companies to expand into the property sector. Taronga's "clean, green and connected" commercial building in Melbourne is using 10 technology solutions which help lower operating costs and at the same time reports a rental premium of 9% with "higher quality tenants" attracted to the buildings sustainability credentials.

Mark Bawtree is CEO at Consulting on Energy, "helping organizations digitalize assets, centralize data, and transform insights into actionable optimization" (Consulting on Energy, n.d.). He emphasises the importance of quality data "PropTech use of AI can be incredibly powerful but it depends on the database. If the data quality is bad then the technology will not function as planned and I have seen this happen. You need the data, including tenant data, to identify where the carbon is. Chief Sustainability Officers are collecting data from multiple sources and in different formats because the smart meters are not in place so the data quality suffers and with the poor data the opportunity to apply AI to decarbonise is being missed".

Despite some doubts about what can actually be achieved, technology has a habit of delivering. Professor Xiaoling Zhang of the Department of Real Estate and Construction at the University of Hong Kong refers to the goals a city like Hong Kong has set itself to be out of reach without adopting technological solutions to decarbonisation. "AI has the potential to map the whole city and provide a tailor made plan with tailor made solutions for every building. The technology makes all the government and business commitments tangible". This is the digital twin.

The digital twin concept dates from 2002 emerging from Product Lifecycle Management and comprises key elements of "real space, virtual space, the link for data flow from real space to virtual space, the link for information flow from virtual space to real space and virtual sub spaces" (Grieves & Vickers, 2106). In the built environment various digital technologies, including Building Information Modelling (BIM), Building Management System (BMS) and Building Automation System (BAS), have converged to enable creation of a digital version of the physical world. Defined by the Centre for Digital Built Britain, a digital twin is "a realistic digital representation of a physical entity". Mark Bawtree defines a digital twin as simply "a digital representation of a physical asset" while again coming back to the importance of the data and notes that it is currently failing to meet expectations because of the poor data quality.

The use of digital twins in promoting a sustainable built environment as:

1. Smart city traffic management, carbon emission reductions, development of sustainable cities.
2. Sustainable design improvement by enabling assessment of alternative designs in the digital world to support decision-making before use in the physical world.
3. Leaner manufacturing process for prefabrication of building components.
4. Construction process optimisation including material use, waste management, machinery use and worker behaviour.
5. Facility management with a digital twin providing a data-driven approach to building operations and management (Shahzad et al., 2022).

The key element of a digital twin, compared to BIM and others, is the ability to take live data feeds from sensors to monitor progress in real time and make adjustments accordingly.

In addition, Professor Zhang imagines community engagement using Augmented and Virtual Reality interfaces to help public understanding of climate related risks and a world where the solutions have been implemented, "climate change is too distant from people's daily lives and struggles", she added. Academic research has used VR to stimulate the experience of homelessness and improve community participation in the response (Alshathri et al., 2024).

The longer term impact of the PropTech contribution to sustainable commercial real estate markets includes:

1. Property management solutions that enable real time monitoring of operations, including energy and water consumption to be compared with sustainability performance targets.
2. Decision-making that is data-driven covering tenant requirements, market trends, operating performance to inform decision-making that helps maximise property value.
3. IoT devices and building technology to improve tenant comfort and experience.
4. Transaction management built on digital platforms that reduce the need for meetings, paperwork and reduce transaction costs.
5. Operating efficiency including energy and water consumption to reduce waste and integrate renewable energy and meet regulatory standards and sustainability reporting requirements (Partners Real Estate, 2024).

Table 10.1 summarises a selection of some of the companies offering sustainability PropTech and some of their technologies and solutions (this is not intended as a comprehensive summary of all companies or all related capabilities or services).

Jonathan Hannam finds it "staggering that we still see many groups that are not making use of technology and that many groups have not changed the way they build and manage assets". Taronga portfolio solutions include Ampd Energy to replace diesel generators on construction sites, Wint software to reduce water consumption (up to 25% savings are claimed). In countries where energy guzzling water desalination plants are used, or water scarce regions, a 25% water saving is significant, "the Gulf Cooperation Council (GCC) countries display a desire to be at the cutting edge of new technology driven sustainability solutions".

As well as voluntary decarbonisation efforts, the need for technology solutions is driven by regulatory requirements and increasing energy costs. In the UK, all private rented homes will be required to have Energy Performance Certificate C or above from 2030 (Department for Energy Security and Net Zero, 2025). According to Saffery a business advisory, tax and accounting firm, PropTech can integrate all the various building energy consuming and generating systems and control lighting, heating,

Table 10.1 Selected PropTech sustainability capabilities

Organisation	*Technology*	*Capability*	*Impact*
Akila	Energy Suite Maintenance Suite Carbon Suite 3D Engine Suite	Digital twin platform to aggregate data inputs from different sources	Decarbonisation through data management and analytics to improve efficiency. ESG reporting
Arup	Climate Risk Screening of Portfolios (CRISP)	Screen property portfolios for carbon-related risks in line with EU regulations	Prioritises carbon reduction measures to protect property values
Aurum PropTech	Over 14 technology products	AI, augmented reality, blockchain, machine learning and virtual reality	Management solutions to save resources and costs
Brainbox	BrainBox AI	AI technology to manage and optimise performance of existing systems	Enable buildings to reduce carbon footprint through energy savings
JLL	Carbon Pathfinder	Assess sustainability data against SBTi targets	Prepare return on investment analysis of decarbonisation options
Proptech OS	RealEstateCore	Open source operating system that uses different applications to gather building data from different silos	Shared language that collates data from different building automation systems and internet of things monitors to optimise energy consumption
Schneider Electric	EcoStruxure Building Platform	IoT-based digital automation and energy management system	Makes all buildings smart to improve efficiency, comfort and productivity
SEKAI	Digital twin technology	Enables real time data transformation into a digital model	A full range of uses including prediction of operational peaks and troughs

(continued)

Table 10.1 (continued)

Organisation	*Technology*	*Capability*	*Impact*
Siemens	Building technology	Energy, HVAC, safety, security, services, digital building life cycle	Building automation, technology and emissions reductions
Taronga Group	Alternative asset manager focused on the intersection of technology and real estate	3 pillars: technology investments, ecosystem support and direct investment in technology enabled real estate including brown to green and "clean, green and connected".	Focus on improving built environments through technology. Direct investment in tech enabled building

(Akila, n.d.; Arup, n.d.; Aurum PropTech, n.d.; BrainboxAI, n.d.; JLL, n.d.; Schneider Electric, n.d.; Sekai, n.d.; Siemens, n.d.; Taronga Group, n.d.)

ventilation and so on as well as providing the data required for sustainability reporting and fostering collaboration between landlord and tenant (Thomas, 2025).

Technology also helps drive efficiency and sustainability in the construction sector through Design for Manufacture and Assembly (DfMA), a combination of Design for Manufacturing and Design for Assembly. This results in building elements being built off-site in a controlled and safe environment to highly standardised designs to be later put together on site. Sustainability benefits of DfMA include reduced waste, less on-site work means energy savings, adoption of other sustainability standards including material selection and greater safety (UCEM, 2025).

Analysis of DfMA in Hong Kong residential construction projects supports the standardisation of modules and spaces for use in different projects and with mass customisation without compromising the ability to deliver varied and creative designs (Laovisutthichai & Lu, 2025). Scaling this up can bring numerous sustainability and economic benefits.

Use of PropTech may increase data security risks, an important consideration for organisations and largely unregulated. Data breaches can be embarrassing and impact reputation. As Bawtree says, "at an operational level, data metrics across a region can be a difficult management issue and like for like analysis within a portfolio is difficult. Energy consumption depends on the outdoor temperature and a number of other variables so the analysis can only really be based on a before and after improvement. At the same time, demands for more data are being made without any regulations in place, resulting in systemic aversion to the data transparency that is needed to drive sustainability action".

Building Occupiers

Environmental considerations for occupiers are related to the impact they make on natural resources, carbon emissions and climate change. Key metrics are electricity consumption from air conditioning, IT, lighting and pantry areas as well as water and waste. Indoor air quality, thermal comfort and natural lighting are also considered. Occupier social considerations might include employee well-being (including wellness and mental health), community engagement and volunteering. When it comes to governance occupiers may be concerned with the ESG reporting process with data inputs from their landlord, data security and upstream and downstream suppliers which includes the landlord and property management teams (Li et al., n.d.). Think of the effort that car manufacturers go to, the number of sensors installed, the personalisation of thermal comfort for passengers in different seats, the air filtration, tyre pressure monitors, auto pilot and the rest. Contrast that with the average office environment and Hannam's point above is well made, it applies to tenants too.

IT Company Case Study

Based on the Google, 2023 Sustainability Report it is possible to identify the following implementation steps and targets that impact their owned and leased offices:

1. Design to international green building standards.
2. Use energy management strategies, building optimisation and data analytics to drive efficiency measures.
3. Circular design and construction by designing adaptable spaces that are less material intensive.
4. Incorporate material reuse in selection process.
5. Prioritise healthy materials that use safer chemicals by use of third-party standards and transparency.

6. Electrification of offices as an essential first step to transition away from fossil fuels and decarbonise operations. Includes electrification requirements in leases, retrofitting owned buildings and electrification standards for new buildings.
7. Battery-based backup power system rather than diesel generators.
8. Energy audits to identify efficiency measures.
9. Measuring refrigerant leaks, preventing leaks and transitioning to low-GWP refrigerants. Patent applied for machine learning technology solution to prevent emissions from leaks in 2022.
10. Drive water use efficiency by various means including design standards, water efficient fixtures and irrigation methods, water leak detection and certification.
11. Zero waste to landfill from data centres.
12. Target 99% diversion of food waste from landfill by composting and other processing means.
13. E-waste recycling.
14. Protect nature and enhance biodiversity by building for ecology, protecting nature and making it accessible, source sustainable materials and support biodiverse food systems, developing technology to reduce biodiversity loss, including AI solutions.
15. Use of native plants in landscaped areas and use bird friendly design to reduce bird strikes on buildings.
16. Engage with suppliers to help them reduce their emissions through a Supplier Code of Conduct.
17. Promote sustainable business travel and commuting practices by providing EV charging stations for 10% of parking spaces at Bay Area headquarters and other new buildings (Google, 2023).

It is also worth noting here the tension between AI as a driver of low carbon, AI related energy consumption challenges and politics. A September 2025 report published in Canada's National Observer notes the ambiguity in company statements citing Google removing a net zero headline from their website while retaining a commitment to net zero by 20,030. Google's total annual energy consumption increased by 26% in 2024 (Snieckus & White, 2025). It seems strategic sustainability statement ambiguity is the means by which various stakeholder threads can pass through the eye of a single needle.

Behaviours are changing According to Stephen White "off-gassing by carpet and paint in particular, provides immediate feedback to building users and tenants". White sees this as one part of wider changes in organisational behaviour by both property developers and tenants. Health concerns about IAQ in offices in China led to increased monitoring, availability of data and awareness. In China's oversupplied office markets from the late 1990's, property developers responded with greater attention to building materials and IAQ. The outbreak of Severe Acute Respiratory Syndrome (SARS) from 2002 to 2004 added to the impetus for sustainable built environments in order to remain competitive. "The game has changed and will remain so in the future for the quality building market", adds White.

Finance Sector Case Study

Another version of this book would have started and finished discussing how green finance is the single most important driver for a sustainable transformation and real estate debt financing. Finance is an essential part of the built environment value chain and credit, or debt, markets are the largest component of the global financial system (Clayton et al., 2021). It's also worth considering the scale of the built environment in terms of property value. According to Savills, the total value of the world's property was US$379.7 trillion by the end of 2022, four times global gross domestic product and more than the global bond and equity markets combined. 13% of this is commercial property and 75% residential (Tostevin & Rushton, 2023). In comparison, total global debt was around US$250 trillion in 2023 (Poplawski, 2024), less than the value of all the property in the world. The market for real estate loans, meanwhile, is estimated to be growing at a compound annual growth rate of 12.1% in 2024 and forecast to be worth US$19.65 trillion in 2029 (Research & Markets, 2025). We know from the global financial crisis that real estate has a tendency to get caught up in the "vicious cycle of value destruction". The words property, debt and crisis seem to be intrinsically linked.

Can we expect a sustainable built environment without sustainable debt? In 2024, the global sustainable finance market was around US$5.9 trillion and cumulative average growth rates (CAGR) of 19.8% between 2025 and 2034 (Global Market Insights, 2025). A significant portion of this will make its way into built environment projects.

Let's start with banking. Rather than a single definition, a range of terms are used to describe sustainable banking, "the combination of strategies and instruments that consistently align daily activities with social, ecological and economic concerns in order to improve the ecological footprint" which requires transparent reporting of non-financial metrics (Riegler, 2023). The UNEP Finance Initiative sets out 6 principles of responsible banking (alignment, impact and target setting, clients and customers, stakeholders, governance and culture, transparency and accountability) (UNEP, n.d.).

Standard Chartered Bank (SCB) is one of three note-issuing banks in Hong Kong SAR and has a history dating back to 1859, according to their website. A broad range of sustainability initiatives cover their client relationships and their business operations. Their published Environmental and Social Risk Management Framework sets out their efforts to achieve a balanced approach to economic, environmental and social needs and the process for identifying, assessing and managing client relationship risks.

The document sets out the governance structure and roles and responsibilities. Part of their sustainability strategy includes public position statements on agribusiness, chemicals and manufacturing, extractive chemicals, infrastructure and transport, power generation, thermal coal, defence and gambling. These set out explicit criteria of activities that must not be undertaken, those that will only be provided for clients that meet certain criteria thresholds, standards expected of clients and best practices that clients are encouraged to undertake. There is a caveat that not all clients will meet all criteria all of the time, including when outside their control (Standard Chartered, n.d.).

When it comes to their property occupancy strategy, sustainability is a core priority for SCB. "The bank is focusing on energy efficiency, waste reduction, and sustainable building practices, ensuring that its real estate portfolio aligns with the company's broader environmental goals" (Cushman & Wakefield, 2024).

The Role of Green Finance

The "UN Principles for Green Financing for Sustainable Real Estate, Infrastructure and Urban Transformation Projects" were published in March 2024. The principles can be implemented at national, regional and local levels through regulatory frameworks and are aimed at promoting the green economy, blue economy and the use of green bonds (Bambagioni, 2024).

The principles are:

1. Early and continuous engagement of public and private capital providers with the planning process, construction companies, developers and other stakeholders.
2. Prioritise real estate an urban transformation projects that aim to:
    a. Reduce climate footprint.
    b. Offer resilience to natural disasters.
    c. Use renewable energy.
    d. Offer affordability and access to all.
3. Green investments should also support housing affordability an urban infrastructure for all through principal adverse indicators (PAI) and social taxonomies.
4. Green development of urban areas to be supported by innovative financial tools (green bonds, green loans, green mortgages, green leases and public–private partnerships) as promoted by government and financiers that enable all stakeholders to contribute.
5. Dedicated bodies shall be established to support development of regulations and oversee green real estate funding and application of Key Performance Indicators (KPIs) to evaluate and assess borrowers delivery of projects.
6. Develop financial and regulatory frameworks to guarantee long-term capital flows to achieve sustainable development objectives, including during economic recession. Design of projects should consider the life cycle and long-term performance.
7. Establish a multi- and cross-sectoral ecosystem of participants to provide long-term support to urban transformation projects (Bambagioni, 2024).

Green finance is good business. In 2023, global banks generated more fees from financing green projects than they did from fossil fuel industries, according to Bloomberg data. A total of US$583 billion in green bonds and loans were made in 2023 with European banks leading the way (Bloomberg, 2024). At the same time, only 7% of green loans and 12% of green bonds in the 5 years up to late 2024 went to real estate projects (Grainger, 2024). JLL report that the retrofitting of the 80% of existing buildings that will still be in use by 2050 need green finance, regulatory support and lenders to fund the work required and avoid the risk of becoming stranded assets.

According to Sunny Zhang, the core appeal of high-performance buildings to green finance lies in their value stability and growth potential. "Relying on the BOMA COE certification system, Shanghai Tower has built a "smart + green" vertical community ecosystem through 21 operational practices that exceed international benchmarks. Its asset risk resistance and long-term return expectations are significantly higher than the market average. This positive cycle of "performance - value - credit" has made it a preferred target for long-term capital such as green bonds and REITs. As revealed by BOMA industry data, the dual support of green certification and efficient operations can not only bring rental premiums but also achieve a substantial increase in capital value, providing a solid value anchor for green finance".

As discussed in the valuation section in Chapter 2, for financial markets, "quantifiable data on the operational side is the key to solving the information asymmetry in green investment; for commercial real estate, green financial capital provides the necessary financial support for operational upgrading. This symbiotic model of "operational practices laying the foundation and financial capital empowering" is not only a win–win path for commercial real estate to achieve carbon neutrality goals and enhance asset value but also the core logic for the real implementation of green finance", Zhang added.

Retrofit Versus Redevelop

Building owners used to focus only on financial, planning and practical options when deciding to retrofit existing buildings or demolish and redevelop. Sustainability adds a newly significant consideration to the decision-making process and conventional wisdom, set out in Table 10.2, suggests that retrofit is always less carbon intensive than redevelopment.

Case studies suggest that things are more nuanced than the table above suggests. A hierarchy of options exists, each with different embodied and operational carbon performance and a range of other factors to consider alongside all the usual planning issues. While retrofitting makes sense in many projects, there are cases where the embodied carbon from redevelopment and retrofitting are only 10% apart. Then it is a relatively simple case of demonstrating how operational carbon can make up the difference over the building lifetime.

Table 10.2 Retrofit vs redevelop carbon savings (Green Construction Board, 2013)

Option	*Carbon saving*	*Comment*
Build nothing	100%	Explore alternatives e.g. Continue to use existing buildings
Build less	80%	Optimise existing building with extension where needed
Low carbon	50%	Use low carbon materials and minimise construction carbon
Build efficiently	20%	Use new construction technology, eliminate waste

According to Fred Pilbrow, Founding Partner of Architects Pilbrow and Partners, the nuance is in "selectively investing carbon in long-term development. This means building with care, as long term stewards of the environment, and doing away with the throw away culture of being careless with carbon". A high-profile case Pilbrow and Partners have been working on is the redevelopment plan of the Marks and Spencer (M&S) flagship building at 458 Oxford Street in London. This is the first major project where embodied carbon has been a significant consideration and much of the debate concerning the retrofit vs redevelop options have been about embodied carbon and calculation methodologies.

One argument is that public amenity value in a new building should be a consideration when determining whether or not carbon-intense redevelopment is worthy of approval or not. The M&S redevelopment will restore a long-lost lane as an arcade, improving porosity to the roads behind Oxford Street and providing a new public space.

As Pilbrow explains, "the M&S proposals at 458 Oxford Street deliver an A1/A5 (materials and construction) embodied carbon level of 651 kg CO2e per sq.m. We have made commitments with M&S to reduce this to 585 kg CO2e per sq.m and we will seek to exceed this target as we develop the detailed design". This compares to estimates of 544 kg CO2e per sq.m for the refurbishment option. Improvements include the use of arched slabs to reduce the amount of concrete by around one-third. Other initiatives include the use of material passports to enable reuse of structural beams at end of life. The aspirational target is for 464 kg CO2e per sq.m

One reason for the extent of the controversy relates to governance of embodied carbon. There are no legislated mandatory standards, creating a considerable grey area. In this void, politicians required to make decisions, may find it safer politically to be a NIMBY (a "not in my back yard" advocate for no development).

How the Empire State Building Set Out to Reduce Energy Consumption by 40%

According to the Clinton Foundation, the Empire State Building retrofit model has been used by more than 50 other buildings worldwide. In 2009, the building, heated by district steam, used as much energy per day as 40,000 single family homes, the plan to reduce energy consumption by 40% involved collaboration by the Empire State Realty Trust (ESRT), the Rocky Mountain Institute, Johnson Controls, Jones Lang LaSalle, C40 Cities Climate Leadership Group and the Clinton Climate Initiative (The Clinton Foundation, 2014).

ESRT has set a target for the building to achieve net zero by implementing a pathway in phases for not only the Empire State Building, but for the entire commercial property portfolio. Combining energy efficiency, renewable energy tariffs, and a 20% residual offset of off-site green energy generation and renewable energy certificates (RECs) is planned to deliver an 80% reduction in operational energy and net zero by 2035 (Retrofit Playbook for Large Buildings, n.d.).

After an initial 200 energy conservation measures (ECMs) were identified and analysed, 60 were selected for implementation in 5 strategically grouped packages over a 15 year period. Presenting a business case involves financial analysis, design analysis, engineering analysis and getting all the moving parts aligned.

Ian Booth is Partner and Decarbonisation Expert at Buro Happold Engineering in New York, he explained that "to maximize the business case we focused on the grouping of energy conservation measures that align with the existing central plant replacement and maintenance schedule. This would include switching from like for like replacement to a solution that considers alternative technologies that can maximize heat recovery or provide both heating and cooling in lieu of heating or cooling only, for example replacing chillers with heat pumps". Booth added, "Very few projects are able to replace ageing chillers early to achieve decarbonisation targets alone. There has to be a business case that considers capital improvements, energy savings and pass through costs to the tenants who benefit from the energy savings. Our focus has been on optimising the design of new tenant spaces, heat recovery and reducing energy consumption through advanced controls and sequences of operation".

As reported in Time Magazine, the CEO of Empire State Realty Anthony Malkin confirmed "there's no silver bullet. It's a silver buckshot. It's lots of pieces that work together". That included complete refurbishment of the chillers, refurbishment of 6,514 windows, installation of regenerative braking on the lifts all saving a reported US$4 million per year (De La Garza, 2021) (Table 10.3).

(Harrington & Carmichael, 2009).

This work has taken place in the context of the New York City Council's Local Law 97 which sets a mandatory requirement for all buildings in excess of 25,000 sq.ft to cut greenhouse gas emissions by 40% before 2030 and 80% before 2050. Ian Booth noted that, despite cynicism in the market, the cost savings achieved have broadly been in line with the expectations above and of note has been the increasing cost of utilities. The use of heat pumps and heat recovery have defied a lot of the conventional wisdom that favoured a business as usual approach. The ability of this work to change the mindset of the designers in the market is what really makes a difference, it is bigger than ESRT we are "educating the market to do better for the greater good while some in the market are doing nothing other than wait for fines while seeing vacancy increase and capital values impacted" added Booth.

According to the case study materials, critical success factors include the energy modelling, identifying the optimal techno-economic package, embracing innovative technology and engineering solutions, incentives and avoidance of future LL97 fines. In this respect "The Empire State Building is as innovative today as it was on the day it was built and serves as the international beacon of possibilities within the built environment to prove the business case to reduce carbon emissions" according to Tony Malkin, Chairman, President and CEO, Empire State Realty Trust.

Financed Emissions

Bank financed emissions can be a source of sustainability tension. Policy Paper 26 published by the Research Centre for Sustainable Hong Kong, City University of Hong Kong, used a top-down approach to estimate emissions from mortgages, business loans and asset management activities in the financial sector of Hong Kong from data published by

Table 10.3 Costs of energy savings at the empire state building

Item	*Cost (US$ million)*	*Description*	*Remark*
Window retrofit	4.5	6,514 windows remanufactured on site to cut winter heat loss by over two thirds and summer heat gain by half Reduced condensation and improved occupier comfort	Annual savings of US$410,000
Radiative Barrier	2.7	Over 6,000 insulating barriers installed on external walls to direct heat into the building	Annual savings of around US$190,000
Demand control ventilation	5.6	Internal CO2 sensors to adjust fresh air mix reducing unnecessary heating / cooling of fresh air	Annual saving of around US$117,000
VAV AHUs	2.4	Replacement of constant volume air handling units to improve efficiency and enhance comfort	
Daylighting, lighting and plug loads	24.5	Lighting upgrades with sensors to make better use of daylight Reduce lighting power density in tenants leased spaces, occupancy sensors for workstations to manage plug load	Annual savings of around US$941,000
Tenant energy management	0.365	Digital dashboard to give tenants real time energy data. Workstation load management and tenant awareness	Annual savings around US$396,000
Chiller plant		Reduce overall building cooling demand to reduce required cooling capacity enabling old chiller plant to be renovated and upgraded rather than replaced. This delivered the savings previously budgeted for	US$14.4 million
Annual energy savings year 1		Cascade of energy savings related to reduced heat gain, radiative barriers to direct heat into the building, daylighting and lighting controls	US$4.4 milion

annual reports of major financial institutions, the Hong Kong Monetary Authority and the Securities and Futures Commission of Hong Kong. This approach suggests an estimated 465 million tonnes of CO2 equivalent in 2021 compared to the 34.7 million tonnes of CO2 equivalent declared by the Hong Kong SAR Environment and Ecology Bureau figure (Dong et al., n.d.). In this case the calculation of financed emissions in Hong Kong is the source of tension, with real estate, professional and business services services together with construction accounting for 13.1% of the Hong Kong economy in 2021 (Census and Statistics Dept, HKSAR Government, 2023) but also being part of other sectors, such as retail, it is easy to see why financed emissions for the built environment make the role of banks so important in driving adaptation and sustainability.

Summary

Sustainability tensions are everywhere. As described above, they can be a driver for positive, innovative, change, and can also be an excuse to delay and continue with business as usual. Where there is doubt, it would be expected that supporting data would be able to persuade even the most reluctant. Even with statistical evidence, denial is sometimes the default response. An example might be "*please share the source of the EUI data quoted. The data and statistics are misleading and inaccurate*". The long reactionary tail of denial on the journey to sustainability performance improvement has yet to be overcome, even when there are obvious financial savings to be made.

As the sample quote above illustrates, organisational dysfunction in the building and construction industry sometimes results from tension in the face of struggles to adapt to the various moving targets in a sector where investment in buildings used to represent relative permanence and financial stability. With new emphasis on ESG reporting there are new things to consider and, like a mature tree with both creaky old branches and flexible new growth, not all stakeholders are moving in the same direction or at the same pace.

In a more positive context, challenge of accepted wisdom is part of the learning process. As the Marks and Spencer building redevelopment shows, retrofits are not always the lowest carbon option. The capacity to understand and meet all these challenges takes a different kind of strategic thinking; when applied effectively sustainability improves financial performance. Finance is still first but this represents a change to the traditional dollars and cents approach, now organisations have to embed sustainability into the finance function.

The sustainability management challenge is well on the way to changing the way the world does real estate in ways we do not yet fully understand. They way organisations respond to sustainability tensions will determine their growth trajectory. Context is critical and technology is a significant part of the solution.

References

Akila. (n.d.). *Akila3D*. Retrieved April 22, 2025, from https://www.akila3d.com/

Alshathri, I., Komlodi, A., & Kuber, R. (2024, October). Investigating the role of virtual reality to support community engagement. *International Journal of Instruction*. https://doi.org/10.29333/iji.2024.17422a

Arup. (n.d.). *Decarbonisation*. Retrieved April 17, 2025, from https://www.arup.com/services/decarbonisation/

Aurum PropTech. (n.d.). *Making you future ready*. Aurum PropTech: A Leading PropTech Company in India. Retrieved April 22, 2025, from https://www.aurumproptech.in/

Bambagioni, G. (2024, April 1). *UN principles for green financing for sustainable real estate, infrastructure and urban transformation projects*. UNECE. Retrieved April 22, 2025, from https://unece.org/sites/default/files/2024-04/01GBambagioni_Green-Financing_20240319.pdf

Bernoville, T. (2020, July 24). *10 rules for a plastic-free office*. Retrieved April 20, 2025, from https://plana.earth/academy/10-rules-plastic-free-office

Bloomberg. (2024, January 4). World's biggest banks made $4 billion on green debt in 2023, beating fossil fuel earnings. *The Straits Times*. https://www.straitstimes.com/business/world-s-biggest-banks-made-4-billion-on-green-debt-in-2023-beating-earnings-from-financing-fossil-fuel)

BrainboxAI. (n.d.). *Welcome to the Green Building Revolution*. Retrieved April 22, 2025, from https://brainboxai.com/en/about

Census and Statistics Dept, HKSAR Government. (2023, June 20). *C&SD: gross domestic product by economic activity*. Census and Statistics Department. Retrieved April 22, 2025, from https://www.censtatd.gov.hk/en/wbr.html?ecode=B10300042023QQ01&scode=250

Clayton, J., Devaney, S., Sayce, S., & van de Wetering, J. (2021). *Climate risk and commercial property values: a review and analysis of the literature*. Retrieved March 8, 2025, from https://www.unepfi.org/wordpress/wp-content/uploads/2021/08/Climate-risk-and-real-estate-value_Aug2021.pdf

Committee on the Financial Aspects of Corporate Governance. (1992, December 1). *Financial aspects of corporate governance*. European Corporate Governance Institute (ECGI). Retrieved April 22, 2025, from https://www.ecgi.global/sites/default/files/codes/documents/cadbury.pdf

Consulting on Energy. (n.d.). *Smart energy choices*. Retrieved April 13, 2025, from https://www.linkedin.com/company/consulting-on-energy/

Cushman & Wakefield. (2024, September 16). *Standard chartered: Transforming a global real estate portfolio.* Retrieved March 14, 2025, from https://www.cushmanwakefield.com/en/insights/standard-chartered-transforming-a-global-real-estate-portfolio

De La Garza, A. (2021, May 10). The Empire State Building's green retrofit was a success. Will other buildings follow suit? *Time*. https://time.com/6026610/empire-state-building-green-retrofit/

Department for Energy Security and Net Zero. (2025, February 7). *Warm homes and cheaper bills as government accelerates Plan for Change*. GOV.UK. Retrieved April 22, 2025, from https://www.gov.uk/government/news/warm-homes-and-cheaper-bills-as-government-accelerates-plan-for-change

Dong, L., Mo, P., Li, L., & Li, K. O. (n.d.). Estimation of Financed Emissions in Hong Kong: From the perspective of bank loans and asset management. *1 City University of Hong Kong Research Centre for Sustainable Hong Kong, Policy Paper 26*. https://www.cityu.edu.hk/cshk/files/PolicyPapers/CSHKPP26_ENG_Final.pdf

Epstein, M. J., Buhovac, A. R., & Yuthas, K. (2015, February). Managing social, environmental and financial performance simultaneously. *Long Range Planning*. https://doi.org/10.1016/j.lrp.2012.11.001

Global Market Insights. (2025, March). *Sustainable finance market size - By investment, by transaction, by end use, by investor, analysis, share, growth forecast, 2025 - 2034*. Retrieved August 15, 2025, from https://www.gminsights.com/industry-analysis/sustainable-finance-market

Google. (2023, July). *2023 environmental report*. Retrieved March 28, 2025, from https://sustainability.google/reports/google-2023-environmental-report/

Grainger, G. (2024, November 15). *Banks and debt providers: The key to unlocking green finance*. The World Economic Forum. Retrieved April 22, 2025, from https://www.weforum.org/stories/2024/11/financial-institutions-can-unlock-green-finance-in-real-estate/

Green Construction Board. (2013, November 25). *Infrastructure carbon review technical report*. Retrieved August 21, 2025, from https://www.constructionleadershipcouncil.co.uk/wp-content/uploads/2019/06/Infrastructure-Carbon-Review-Technical-Report-25-11-13.pdf

Grieves, M., & Vickers, J. (2106). Origins of the digital twin concept. *Working Paper*. https://doi.org/10.13140/RG.2.2.26367.61609

Hansor, R. (2025). *Deloitte scope 3 community webinar*. Retrieved March 27, 2025, from https://www.linkedin.com/posts/roberthansor_sustainability-scope3-netzero-activity-7309921543109726209-TxWj/

Harrington, E., & Carmichael, C. (2009). *Project case study: Empire State Building*. Retrieved March 6, 2025, from https://rmi.org/wp-content/uploads/2017/04/Buildings_Retrofit_EmpireStateBuilding_CaseStudy_2009.pdf

IEA. (2024, October 9). *Massive global growth of renewables to 2030 is set to match entire power capacity of major economies today, moving world closer to tripling goal - News*. IEA. Retrieved July 8, 2025, from https://www.iea.org/news/massive-global-growth-of-renewables-to-2030-is-set-to-match-entire-power-capacity-of-major-economies-today-moving-world-closer-to-tripling-goal

JLL. (n.d.). *Decarbonisation strategy*. Retrieved April 22, 2025, from https://www.jll.com.hk/en/solutions/decarbonization-solutions

Kell, G., Reeves, M., & Fox, H. (2025, April 22). *Corporate sustainability is in crisis. What should companies do now?* Harvard Business Review. Retrieved April 25, 2025, from https://hbr.org/2025/04/corporate-sustainability-is-in-crisis-what-should-companies-do-now

KPMG Australia. (2024, June 27). Retrieved March 8, 2025, from https://kpmg.com/au/en/home/insights/2024/06/kpmg-esg-assurance-maturity-index.html

Laovisutthichai, V., & Lu, W. (2025, January 29). Design for manufacture and assembly (DfMA) for modular buildings: an analytical framework. *Journal of Asian Architecture and Building Engineering*. https://doi.org/10.1080/13467581.2025.2455028

Li, C., Kay, L. W., & How, P. (n.d.). *Why ESG matters: A guide for occupiers*. Retrieved April 5, 2025, from https://content.knightfrank.com/resources/knightfrank.com/expo/esg-checklist-for-occupiers-research-paper.pdf

Partners Real Estate. (2024, June 7). *How PropTech is changing the commercial real estate landscape partners real estate*. Partners Real Estate. Retrieved April 22, 2025, from https://partnersrealestate.com/media-center/blog/how-proptech-is-changing-the-commercial-real-estate-landscape/

Poplawski, M. (2024, December). *Global Debt Monitor 2024*. International Monetary Fund (IMF). Retrieved April 22, 2025, from https://www.imf.org/external/datamapper/GDD/2024%20Global%20Debt%20Monitor.pdf

Proptechbuzz. (2025, March 27). *Six PropTech Solutions: Transforming Real Estate Sustainability Challenges into Opportunities*. Retrieved April 8, 2025, from https://www.proptechbuzz.com/blog/six-proptech-solutions-sustainability

Research and Markets. (2025). *Real estate loan market report 2025*. Retrieved March 4, 2025, from https://www.researchandmarkets.com/report/real-estate-loans-market

Retrofit Playbook for Large Buildings. (n.d.). *New York City icon reaches for carbon neutrality*. Retrieved September 29, 2025, from https://retrofitplaybook.org/resource/empire-state-building/

Riegler, M. (2023, March 31). Towards a definition of sustainable banking - A consolidated approach in the context of guidelines and strategies. *International Journal of Corporate Social Responsibility*, *8*. https://doi.org/10.1186/s40991-023-00078-4

Schneider Electric. (n.d.). *Enable smart buildings with EcoStruxure Building*. Explore EcoStruxure Building. Retrieved April 22, 2025, from https://www.se.com/hk/en/work/campaign/innovation/buildings.jsp

Sekai. (n.d.). *Engineering at the essence*. Retrieved April 22, 2025, from https://www.sekai.io/about

Sellers, K. (2022). *PropTech: Solutions for sustainability and cost mitigation*. Retrieved March 8, 2025, from https://www.savills.com/impacts/technology/proptech-solutions-for-sustainability-and-cost-mitigation.html

Shahzad, M., Shafiq, M. T., Douglas, D., & Kassem, M. (2022, January 25). Digital twins in built environments: An investigation of the characteristics, applications, and challenges. *Buildings*, *12*(2). https://doi.org/10.3390/buildings12020120

Siemens. (n.d.). *Building technology*. Siemens. Retrieved April 22, 2025, from https://www.siemens.com/hk/en/products/buildingtechnologies.html

Snieckus, D., & White, R., (2025, September 4). *Google deletes net-zero pledge from sustainability website*. Canada's National Observer. Retrieved September 17, 2025, from https://www.nationalobserver.com/2025/09/04/investigations/google-net-zero-sustainability

Standard Chartered. (n.d.). *Environmental and social risk management framework*. Retrieved March 12, 2025, from https://av.sc.com/corp-en/nr/content/docs/esrm-framework.pdf

Taronga Group. (n.d.). *At the forefront of innovation*. Retrieved April 22, 2025, from https://tarongagroup.com/about-us/

The Clinton Foundation. (2014, August 13). *Four reasons why the empire state building retrofit model works*. The Clinton Foundation. Retrieved April 22, 2025, from https://stories.clintonfoundation.org/four-reasons-why-the-empire-state-building-retrofit-model-works-1e4d9ebdbe9

Thomas, Z. (2025, March 18). *PropTech: Supporting sustainability ambitions in real estate*. Retrieved 12, April, from https://www.saffery.com/insights/articles/proptech-supporting-sustainability-ambitions-in-real-estate/

Tostevin, P., & Rushton, C. (2023, September 20). *Total value of global real estate: Property remains the world's biggest store of wealth*. Savills. Retrieved April 22, 2025, from https://www.savills.com/impacts/market-trends/the-total-value-of-global-real-estate-property-remains-the-worlds-biggest-store-of-wealth.html

UCEM. (2025, March 21). *What is DfMA (and what does it mean for construction)?* UCEM. Retrieved April 22, 2025, from https://www.ucem.ac.uk/whats-happening/articles/what-is-dfma/

UNEP. (n.d.). *Creating the future of banking*. Retrieved March 7, 2025, from https://www.unepfi.org/banking/more-about-the-principles/

Van der Byl, C., & Slawinski, N. (2015). Embracing tensions in corporate sustainability: A review of research from win-wins and trade-offs to paradoxes and beyond. *Organisation and Environment*, *28*(1). https://doi.org/10.1177/1086026615575047

WBCSD. (2020). *Board directors' duties and ESG considerations in decision-making*. https://archive.wbcsd.org/contentwbc/download/10820/160014/1

CHAPTER 11

Lessons from Other Industries

"When you drop any new idea in the pond of the world, you get a ripple effect. You have to be aware that you will be creating a cascade of change."
Joel A Barker

Industry and Sustainability

Agribusiness is a useful case study to start with. It is easy to understand the inputs and outputs, including labour, land, water, processing, packaging and transportation. The monitoring of data seems relatively straightforward and can be clearly measured and communicated. We look at sustainable wine production in New Zealand; this is a model for the world.

Aviation is a high-profile energy intensive sector. It delivers the pollution directly into the upper atmosphere; most of us will be well aware of the waste generated and have perhaps observed the changes in the industry post-COVID-19. For example, Air China launched its Clean Plate Campaign in line with government concerns about food security during the pandemic (BBC, 2020). More significantly the industry around the world is taking steps to transition to sustainable aviation fuel.

Another key sector is shipping. Driven by the need to protect expensive ships, valuable cargo and meet crew requirements for safety, the maritime

S. Crispin, *Sustainability in the Built Environment*,
https://doi.org/10.1007/978-981-95-6549-8_11

industry has a long history of setting standards for navigation and "rules of the road". The British Steam Navigation Act of 1846 mandated red for port and green for starboard. Within 12 years Prussia, France, Austria and Hungary had all adopted the same standard. This later became part of the International Regulations for Preventing Collisions at Sea (Tibke, n.d.). The Merchant Shipping Act of 1876 mandated load lines (the Plimsoll Line) to avoid overloading vessels.

The footwear industry is a complex challenge for good environmental practice. Each year an estimated 300 million shoes are sold, thrown away and end up in landfill. The biggest driver for sustainability is the expectations of a younger generation of environmentally conscious consumers; they want a sustainable shoe that combines fashion and awareness. Even US Presidential candidates have been seen flaunting the "political shoe" (Hernandez, 2024). At the bottom of the socio-economic pyramid, footwear is an item that many people struggle to afford at all. There are no easy answers to unpicking the mix of leather, petrochemical derived materials, metals and adhesives that go into this "everyday" item. It is variously status symbol, fashion statement, practical item, protective clothing and social differentiator.

Aquaculture is another agribusiness and source of much controversy. One only needs to recall the anti-whaling protests or the news about plastics dumped at sea to know how deeply the oceans can stir emotions. Aquaculture is sea and land based and there are movements to find sustainable farming models that help deliver food security to the poor and fine dining with a conscience to the more fortunate. Land-based fish farming offers some interesting insights into the role of technology in delivering sustainable solutions.

Wine Making

Wine making in New Zealand is a sustainability model for the world. By 2022, 96% of New Zealand vineyards were certified sustainable, over 90% of wineries and vineyards have water conservation plans and 98% have recycling or waste reduction plans (Pomranz, 2022).

The transition of New Zealand wine production to sustainable practices makes a good starting point. Viticulture is easy to visualise and the benefits of sustainability are clear and obvious. The main inputs are water, soil, fertiliser and pest control sprays, energy (diesel and electricity) and labour. Outputs include grapes, gas emissions to air, vineyard posts and wastewater. The wine production stage also produces grape marc (solid remains of the grape after pressing). Lightweight bottles, solar panels and on-site composting are all used to help reduce the carbon footprint.

Neto et al. (2012), used Life Cycle Assessment to identify all the key input and output categories in Portuguese viticulture as follows:

Inputs:

1. Water
2. Land
3. Phytosanitary products
4. Synthetic fertilisers
5. Solid manure
6. Transportation

Outputs:

1. Grapes
2. Emissions to air from fertilisers
3. Emissions to air from phytosanitary products (Neto et al., 2012)

In 1995, Sustainable Winegrowing New Zealand (SWNZ) developed a certification programme to provide a best practice model for the industry. Drivers from within the industry itself and from customers in export markets develop guidelines for sustainable viticulture and a platform for knowledge sharing. Certification of a vineyard follows the self-assessment using questionnaires, data collection for all inputs and outputs and a 3rd party audit process. This covers major and minor areas and includes spray diaries that covers use of agrichemicals (New Zealand Wine, n.d.an.d.a).

Benefits from participation in the SWNZ programme include:

1. Protecting the environment while efficiently and economically producing a premium product.
2. Improving operations under a continuous improvement programme.
3. Technology and knowledge transfers.
4. Fully sustainable certification programme based on rigorous external audits.

5. An approved industry audited sustainability programme. (Discover Sustainable Wine, n.d.)

This result makes SWNZ one of the leading sustainability programmes in the world over 90% of New Zealand wine is now processed in certified facilities. The certification programme covers the entire supply chain including vineyards, bottling and wineries. Benefits to producers include:

1. Producer sustainability claims supported by a robust framework.
2. Trust of sustainability-aware consumers worldwide.
3. A network of like-minded buyers and sellers of sustainably produced grapes.
4. Quantitative data backed sustainability metrics to help avoid introduction of mandatory regulatory requirements.
5. Data-driven improvement plans to maximise economic and sustainability benefits.
6. Support for continuous improvements.
7. Knowledge sharing (New Zealand Wine. n.d.b).

There is more to be done; the industry has ambitious targets by 2050, including carbon neutrality, zero waste to landfill and other circular industry ambitions (Green, 2022).

New Zealand wine is an example of one industry coming together at a national level to organise itself to make environmental improvements and also market a product. The timing coincided with the "100% Pure New Zealand" tourism industry campaign which featured sustainability attributes such as fresh air, a clean environment and hospitable people. The tourism case was controversial, however, because New Zealand had failed to keep its commitment under the Kyoto Protocol according to The Economist (The Economist, 2010) while the New York Times highlighted the pressures from the dairy industry on the land (Larsen, 2019).

As previously discussed, sustainability is never far from tension and controversy; the balance between financial returns and sustainability performance is difficult to get right. But controversy can also have unexpected benefits. The poster child for turning sustainability controversy into a business case is the Volkswagen (VW) emissions scandal. Cheating emissions tests resulted in damages and fines of over US$30 billion after discovery by the US Environmental Protection Agency (EPA). The resulting formation of a Sustainability Council by VW resulted in a number of changes, technology, policy and culture. The end result was a corporate commitment to decarbonisation and rebranding to become a leader in sustainability (Kell, 2022). Another result of the scandal was changing consumer attitudes towards diesel vehicles as a whole; this prepared the ground for the transition to electric vehicles.

What we learn from these cases is that controversy can sometimes be the quickest route to consumer awareness and changing attitudes which drive deeper sustainability action. We can also find lessons in integration of sustainability targets from two overlapping industries to help each other achieve an even greater good that might translate into lessons for construction and property investment. For highly complex industry supply chains, such as the built environment, this is a source of hope. We can consider how building materials, construction and property management can all come together and achieve more than as separate industries.

Aviation

Aviation is another model. It appears to be a more obvious polluting industry than buildings. But in fact it only accounted for 2% of global CO2 emissions in 2022, even with demand growing again quickly after the COVID-19 pandemic (IEA, 2023). The inputs and outputs include the plane from mineral extraction to end of life, all the components, including navigation systems, engines, oils, batteries, systems, manufacturing, assembly and transportation. At the operation stage of the life cycle are fuel extraction, transportation and processing, airport operations, fuel consumption, and maintenance. At end of life, decommissioning includes disassembly, recycling and waste treatment (Rupcic et al., 2023).

The global industry meets collectively under the co-ordination of the International Civil Aviation Organisation (ICAO), a specialised agency of the United Nations that evolved out of the 1944 Chicago Convention. The core function of the ICAO is to "support diplomatic interactions, and to research new air transport policy and standardization innovations" (ICAO, n.d.). It recognises the primacy of national regulation in aviation and national sovereignty over airspace. ICAO convenes panels, task forces, conferences and seminars which then result in a diplomatic mission to establish international standards and recommended practices for civil aviation. In October 2022, 184 nations met to agree on a range of measures to reduce CO2 emissions in October 2022 and adopted the goal of reaching net zero-carbon emissions by 2050 as a Long-Term Aspirational Goal (LTAG) (WEF, 2022).

The key pillar to achieving this is development of sustainable aviation fuels (SAF) which is estimated to deliver 65% of the emissions reductions needed. While a massive increase in production of SAF is necessary, the current aircraft and fuelling infrastructure in place is compatible with the transition. The balance will be achieved through offsetting under the Carbon Offsetting and Reduction Scheme for International Aviation (CORSIA). The ambitious targets are set against a 2019 baseline and a set of environmental integrity criteria and requirements have been set around offsetting. Exemptions apply for Least Developed Countries, Small Island Developing States and Landlocked Developing Countries (IATA, 2023). In Singapore, the Ministry of Transport announced a mandatory use of 1% minimum SAF for all departing flights from 2026. This is alongside other initiatives to develop an SAF ecosystem.

The ambition, framework, phased approach and exemptions from the aviation industry case give a useful indication of what can be achieved by an energy intense industry. The role of the various industry bodies and the United Nations in bringing the industry together is a demonstration of how such ambitious targets can be set. Difficulties remain of course, but seeing this industry come together and agree to take collective action is a significant example to other industries.

Maritime Shipping

Shipping is another industry that has a lot to learn from. While the maritime industry accounts for around 80% of cargo volume and 70% of cargo value (Vakili et al., 2023), the global shipping fleet accounted for only around 3.3% of global emissions in 2007. Ship exhaust accounted for most of the emissions and pollutants (Faber et al., 2012), often delivered to the densely populated hearts of international port cities as well as sensitive marine environments.

Driven by commercial interests, the maritime industry has long had an international body to bring stakeholders together and implement safety, navigation rules and environmental standards. Under the United Nations, the International Maritime Organization (IMO) was originally known as the Inter-Governmental Maritime Consultative Organization, established at an international conference in 1948. It became the International Maritime Organization in 1982. The original purpose included "prevention and control of marine pollution from ships" and now has the "responsibility for the safety and security of shipping and the prevention of marine and atmospheric pollution by ships" (IMO, n.d.).

Even with a long established effective governing organisation in place, it still took the IMO 10 years to bring stakeholders together to deliver industry-wide alignment to reduce the sulphur content of bunker fuel from 3.5% to 0.5% and agree non-compliance penalties (Harvey, 2020).

Under the Sustainable Development Goal (SDG) framework, the IMO has a comprehensive blueprint for sustainability and a healthier planet. Examples include SDG 7 Affordable and Clean Energy with action on international cooperation for clean energy research and technology. GHG reduction initiatives include:

1. Integrated Technical Cooperation Programme
2. Multi-donor trust fund
3. Global Maritime Technologies Cooperation Centres
4. Green Voyage 2050
5. GHG-Smart Programme and Future Fuels and Technology for Low- and Zero-carbon Shipping Projects
6. The UNDP-GEF GloFouling Partnerships project
7. Transfer of Environmentally Sound Technologies (TEST) Biofouling project
8. IMO Coordinated Actions to Reduce Emissions from Shipping (CARES)
9. The IMO UNEP-Norway Innovation Forum
10. The IMO-EBRD-World Bank Financing Sustainable Maritime Transport Roundtable
11. NextGEN (Green and Efficient Navigation) portal and NextGEN Connect projects

There is also the Capacity Building for Climate Mitigation in the Maritime Shipping Industry to enable the Least Developed Countries and Small Island Developing States to implement energy efficiency (IMO, n.d.).

Further decarbonisation efforts of the industry include the Energy Efficiency Existing Ship Index (EEXI) and the Carbon Intensity Indicator (CII), both adopted by the IMO (Vakili et al., 2023). Other efforts include reducing speeds, or slow steaming. Emissions and fuel consumption decrease exponentially as ship speeds are reduced (Faber et al., 2012) as well as being an effective means for the industry to reduce capacity at times of lower demand, according to The Baltic and International Maritime Council (BIMCO) an industry organisation representing around 60% of shipowners, brokers, agents and charterers (Rasmussen, 2023).

International regulations around slow steaming include global treaties under the IMO, targets to reduce black carbon in the Arctic and local conditions of entry to specific ports. But slow steaming is not a silver bullet; below certain speeds, emissions can increase again with ships and their systems operating at sub-optimal levels (Bartlett, 2023). Reflecting the complex nature of sustainable shipping, rather than a fixed speed limit, a limitation of 75% of a ships design speed have been suggested, resulting in a differentiated restriction for different categories of ships (Buitendijk, 2022).

In common with the built environment, the variables of the maritime industry are as numerous as the complexities. Congested supply chains can result in increased emissions. In 2021, The Port of Los Angeles reported significant increases over 2020, including a 145% increase in sulphur oxides, partly as a result of port congestion. Part of the solution is port optimisation to avoid backlogs of ships at anchor waiting to dock with their engines running (The Port of Los Angeles, 2022). There are parallels to be drawn with building management systems (BMS).

Footwear

The footwear industry represents a great challenge for environmental best practice. Like real estate, it is an everyday "essential" that many have to do without. Around 25 billion sports shoes are sold each year across the globe (Cheah et al., 2013) and an estimated 300 million shoes end up in landfill; rubber soles take 80 years to decompose and the springy midsole of a running shoe, usually made from Ethylene Vinyl Acetate, can kick around in landfill for hundreds of years. There is no global industry association for the shoe industry, but the Footwear Distributors and Retailers of America (FDRA) dates back to 1944; it is a trade organisation for the footwear industry with 500 companies and brands as members, representing 95% of US shoe sales (FDRA, n.d.a).

When it comes to shoes, it may seem that there are no good material options. The inputs of a shoe depend on the material choices such as leather or synthetic upper, rubber or synthetic sole and any number of design variables including metal or stitched eyelets, mid sole cushioning and so on. Leather requires cattle farming, tanning and further processing; synthetic materials are derived from petrochemicals; rubber is generally produced in monoculture plantations carved out of rain forests and the processing requires chemical additives. Shoe production is Asia centric (88% of production) and consumption is US and Europe centric. According to the World Footwear Yearbook 2021, around 13 billion pairs of shoes were shipped in 2021 (World Footwear, 2022).

The footwear industry is changing not because of a UN led global industry body, but because consumers and manufacturers themselves are driving change. Consumers for sports shoes tend to be younger and want to know that they are wearing recycled materials in their shoes. They understand the concept of circularity and want to know that take-back schemes help to close the circular economy loop. While shoes cannot easily be disassembled for direct recycling, they can be ground down and transformed into new products, such as all-weather sports surfaces. The FDRA example shows how national industry associations can enable manufacturers to share knowledge and foster collaboration. It does so through "shoesustainability.com" which is a site to "educate, empower and activate the footwear industry" (FDRA, n.d.b). Global shoe brands have adopted sustainability strategies and are working across the supply chain to re-engineer shoe design and manufacturing processes.

Vinita Turakhia is a champion of technology in advancing the SDGs in both sustainable fashion and well-being. She shared the following thoughts: "When analyzing consumer-led change, it is important to consider regional variations. In the Western world and developed economies, purchasing decisions are increasingly influenced by factors such as the product's origin, manufacturing processes, and environmental impact. However, in developing economies, buying decisions are still primarily driven by the pursuit of value, aiming to obtain high-quality products at the lowest possible cost. Consequently, within the fashion industry, the shift towards sustainability is primarily propelled by a small segment of conscious consumers and responsible investors. Yet, achieving the necessary scale of sustainable solutions to make a meaningful impact remains a challenge.

Manufacturing practices still revolve around large minimum order quantities (MOQs), leading to sales strategies that heavily rely on high discounts to entice customers to buy more. To achieve impactful change, a collaborative effort is required from manufacturers, industry associations and consumer groups. In developing economies, waiting for consumers to voluntarily adopt conscious consumption patterns may result in missed opportunities for action. In such cases, government intervention in the form of policy directives, combined with manufacturers' collective efforts to implement environmentally friendly practices, becomes essential.

While precise data on consumer behaviour and market dynamics across regions would provide further insights, it is evident that a comprehensive and coordinated approach involving multiple stakeholders is vital to drive the necessary sustainability transformations in the fashion industry. In my opinion it is a long way ahead before we see meaningful impact in this sector".

The footwear industry is similar to the built environment in that there is no long history of a global industry association to set rules and boundaries. It is largely up to individual companies to take action and to comply with environmental legislation. As in the aviation and maritime sectors, there are a vast range of brands, manufacturers, suppliers and a complex mix of materials. Historically, sustainability was not a concern. Like buildings, they are an essential and there were few environmental concerns before the sustainability imperative.

Fisheries and Aquaculture

For our final case study, let us return to food production. Anyone who has seen the documentary Seaspiracy will know that the commercial wild-catch fishing industry is controversial enough to suggest we give up eating fish altogether (Read, 2021).

Overfishing of wild fish, controversy around certification systems, self-certification and bycatch are just some of the concerns. On the other hand, fisheries and aquaculture play a significant role in delivering global food security, a healthy protein source and jobs (UNEP, 2014). In some respects this is similar role to that played by the built environment—it meets the basic need for shelter and physical security while also having a significantly detrimental environmental impact. Nearly 60% of protein consumed in Ghana is fish protein (Thiao & Bunting, 2022).

Like all the other cases, developing sustainable land-based aquaculture relies on a range of technological innovations. They are necessary to create complex artificial ecosystems that balance and accommodate an almost infinite range of variables such as water temperature, water chemistry, fish species, feed type, feeding habits (bottom feeders, suspended feeders and floating feeders) and so on. Solutions include developing new plant or insect-based feeds, recirculating water in land-based fish farms (recirculating aquaculture systems or "RAS"), driving energy efficiency and workers' rights and equity. Traditional fish farming has a particular problem with feed produced from fishmeal (smaller fish not generally used for human consumption which are caught and ground up for fish food in commercial farming operations). This depletes the food chain in natural habitats.

Particular problems are that small scale fishers account for 40% of the global catch and 90% of the people that work in the value chain (FAO, n.d.). Offering sustainable solutions to the industry involves reaching a diverse range of individuals and organisations across the globe, a situation that has similarities to the global construction and real estate industry. Another similarity is the labour issues.

Sustainable aquaculture and the sustainable built environment intersect in vertical fish farms using RAS technology. One such company is Vertical Oceans, a Singapore-based start-up that produces "The World's Best Tasting, Most Sustainable Prawns" (Vertical Oceans, n.d.) and part of the Singapore Government "30 by 30 strategy" to produce 30% of nutritional needs by 2030 by 2030 (Singapore Food Agency, n.d.). In the Vertical Oceans closed-loop system, the prawns sit at the top and drive the revenues. Water is then circulated into tanks containing tilapia, a resilient fish that survive in a wide range of waters and help absorb waste from the shrimp tanks, before moving on to the algae stage producing sea grapes, a sashimi garnish.

Key Takeaways

From the different case studies above, we can identify common success factors:

1. A multinational industry body.

2. Cross-border cooperation and harmonisation of standards.
3. A business case for transformation.
4. Technological innovation.
5. Market leaders of sufficient scale to influence the industry as a whole.

This process is readily transferable to local real estate owners associations. The question is one of mainstreaming and gaining sufficient numbers of building owners to commit to a common standard before more onerous regulations are brought in at city level. Indeed, the role of city policy makers is as a catalyst for action that results in a cascade of interventions or a virtuous cascade triggered by a "tipping point". Considering the built environment as an economic system with social system impacts is helpful in preparing such a roadmap (Times Colonist, 2020).

We have already touched on controversy in sustainability and we conclude this chapter by taking a look at how a deeply controversial energy source plots a pathway to scaleable commercialisation. A paper published by the Cascade Institute presents the case for Canada's potential to be a global leader in deep (>5 km below the Earth's surface) geothermal energy generation as a means to bridge the net zero electricity gap (Graham et al., 2022).

What this paper does usefully share is a roadmap to scaleable commercialisation, that could apply to the built environment globally:

1. Data to assess potential scale and location of the opportunity including required well depths in each location.
2. National agency to collect data and facilitate knowledge sharing.
3. Use technology to identify success factors that improve exploration success rates.
4. Identify costs and benefits of different hard rock drilling technology.
5. Low carbon drilling technology to maximise "energy return on energy investment" (EROEI).
6. Determine cost competitiveness under different scenarios.
7. Improve drilling technology and lower costs.
8. Improve reliability and prevent leaks.
9. Investment modelling to achieve a market breakthrough.
10. Model maintenance costs and benchmark against that of other renewable electricity generation.

11. Prepare a national strategy for investment and incentives.
12. Government backed financing schemes (long-term bonds).
13. National Corporation to lead R&D and commercialisation.
14. Consider de-risking options for investments to prepare for commercialisation.
15. Economic impact assessment.
16. Export market potential.
17. Intellectual property ownership. (Graham et al., 2022).

Lessons for the Built Environment

Interesting as these industry cases are, they have little or nothing to do with the built environment. They do demonstrate that a "can do" attitude helps drive innovation and technological advances to implement sustainability, even in some unlikely areas. The key question is what can we borrow and apply to the built environment?

The shipping industry provides the most workable example of a United Nations led industry organisation that brings national governments, industry and suppliers together to lead and drive change. Shipping is by nature a cross-border industry but buildings and cities are not. Property developers, construction companies, labour, material suppliers and even regulators are very local. In China, regulations are enforced at city district level in Australia planning is approved at local council level. Are there simply too many players for a single built environment industry organisation to handle?

We can see from the footwear industry that demand driven change can happen with industry organisations. The FDRA in the United States is a national industry organisation representing 500 companies and brands accounting for 95% of US footwear sales. How much progress is really being made? Their own survey in 2022 revealed that nearly 95% of industry insiders think that little significant or no real progress has been made (FDRA, 2022). How much of this represents the US model of prioritising profit over sustainability? With change happening in the bigger brands, are we seeing a younger generation of consumers driving change in the footwear industry?

Some Positive Vibes

SWNZ may be considered one of the most successful case studies—but is it scaleable? According to Statista, the total revenues from New Zealand wine was US$1254 million in 2022 (Statista, n.d.) with 739 wineries listed in The New Zealand Wine Directory in 2023 (NZ Wine Directory, n.d.). It is a tight and contained market geography with deeply interwoven supply chains and interdependent producers in a nation of reputable agribusiness dependent on branding to attract buyers from remote markets. Can the built environment reach the same wide adoption of sustainable practices.

A positive model comes from the maritime industry, a case study that the built environment can borrow many positive insights from. In particular, the role of the IMO in driving change across a range of stakeholders.

Development of sustainable aviation fuel is broadly similar to the issues of energy use in the hard to abate building material manufacturing. Progress here will be painfully slow but inter-industry learning will help both overcome the challenges and it is positive to see so much collaboration.

The role of RAS technology in aquaculture is one of the more surprising lessons for the built environment. Wide adoption of new technology in both is imperative. Depletion of natural resources, fewer wild fish available to catch, is a key driver in this instance.

Alternative materials in the footwear industry also has parallels with material substitution in construction. Still not widespread, particularly in the lower end of the market, this too is something to watch for the future.

Summary

Despite some parallels, the global property market is a very different beast to all the industries described above. According to Savills the global real estate industry value is nearly US$380 trillion in 2023 (Savills, 2023). China alone had 103,000 real estate development companies in 2022 (Zhang, 2023) and the US over 63,000 housing developers in 2023 (IBIS World, n.d.). We are not going to get many of them around the same table. Trying to agree common sustainability targets will be a thankless task, but we may start to see positives that result from our friend, controversy.

There is a strong business driver. Above all, in the race to net zero as adopted by many companies, countries and cities will start to see decarbonisation as a competitive factor. In Asia, the Singapore and Hong Kong rivalry is well known and the potential to support corporate net zero targets between cities may become a new element in city competitiveness. It is not a matter of either or, but balancing different factors in allocating resources between city A and city B in a scorecard approach, in the same way that clean air, public transport are already a quantified factor in city competitiveness studies (Mori Foundation, n.d.).

The industry organisation approach to establishing common sustainability targets in the built environment is a tough one. There is no readily applicable model but there are lessons to learn from each of these cases—cross-border, consumer driven, solving complex problems with technology and setting aspirational goals.

Transforming the built environment is not just about new green buildings, nor is it about retrofitting existing buildings. The need is to transform the whole built environment over its entire life cycle, from raw material extraction to end-of-life circularity and waste disposal. That does not mean eliminating all CO2 emissions, it means taking all possible measures to minimise them over time. This starts with a focus on commercial property, offices, industrial and retail, then moves to focus on to include residential. Once that is done, provision can be made to remove the residual CO2 emissions to achieve a balance. As we have seen in the previous chapters, so much of the research and modelling has been done and the business case is solid. In the final chapter, we will look at the next steps in the context of reasons to be cheerful.

References

Bartlett, C. (2023, May 6). *Cleaner shipping should be powered by port optimisation, says PortXChange*. The Loadstar. Retrieved February 28, 2024, from https://theloadstar.com/cleaner-shipping-should-be-powered-by-port-optimisation-says-portxchange

BBC. (2020, August 12). *China launches 'Clean Plate' campaign against food waste*. BBC. Retrieved February 28, 2024, from https://www.bbc.com/news/world-asia-china-53761295

Buitendijk, M. (2022, October 31). *IFAW argues for speed limit for ships in EU waters*. SWZ|Maritime. Retrieved February 28, 2024, from https://swzmaritime.nl/news/2022/10/31/ifaw-argues-for-speed-limit-for-ships-in-eu-waters/

Cheah, L., Ciceri, N., Olivetti, E., Matsumura, S., Forterre, D., Roth, R., & Kirchain, R. (2013, April). Manufacturing-focused emissions reductions in footwear production. *Journal of Cleaner Production, 44*. https://doi.org/10.1016/j.jclepro.2012.11.037

Discover Sustainable Wine. (n.d.). *New Zealand*. Discover Sustainable Wine. Retrieved May 19, 2023, from https://discoversustainablewine.com/new-zealand/

Faber, J., Nelissen, D., Hon, G., Wang, H., & Tsimplis, M. (2012). *Regulated slow steaming in maritime transport*. Retrieved February 28, 2024, from https://theicct.org/wp-content/uploads/2021/06/CEDelft_slow_steaming_2012.pdf

FAO. (n.d.). *Home | Food and agriculture organization of the United Nations | International Year of Artisanal Fisheries and Aquaculture*. Food and Agriculture Organization. Retrieved May 30, 2023, from https://www.fao.org/artisanal-fisheries-aquaculture-2022/home/en/

FDRA. (n.d.a). *About us*. FDRA. Retrieved May 30, 2023, from https://fdra.org/about-fdra/

FDRA. (n.d.b). *Shoe sustainability*. Shoe Sustainability. Retrieved May 30, 2023, from https://www.shoesustainability.com/

FDRA. (2022). *2022 Shoe sustainability benchmark & progress report*. Retrieved February 28, 2024, from https://fdra.org/wp-content/uploads/2022/11/FDRA-Shoe-Sustainability-Benchmark-Progress-Report_1022.pdf

Graham, I., Quigley, E., Janzwood, S., & Homer-Dixon, T. (2022, May 27). *Deep Geothermal Superpower Canada's potential for a breakthrough in enhanced geothermal systems*. Cascade Institute. Retrieved February 28, 2024, from https://cascadeinstitute.org/wp-content/uploads/2022/03/Deep-Geothermal-Superpower-version-2.0-.pdf

Green, M. (2022, April 22). *New Zealand vineyards work towards achieving carbon neutrality*. https://www.decanter.com/wine-news/new-zealand-vineyards-work-towards-achieving-carbon-neutrality-478910/. Retrieved February 28, 2024, from https://www.decanter.com/wine-news/new-zealand-vineyards-work-towards-achieving-carbon-neutrality-478910

Harvey, F. (2020, January 1). Shipping fuel regulation to cut sulphur levels comes into force. *The Guardian*. Retrieved May 30, 2023, from https://www.theguardian.com/environment/2020/jan/01/shipping-fuel-regulation-to-cut-sulphur-levels-comes-into-force

Hernandez, J. (2024, February 19). *Trump debuts line of $399 sneakers, one day after $355 million fraud judgment*. NPR. Retrieved February 28, 2024, from https://www.npr.org/2024/02/19/1232438349/donald-trump-golden-high-top-sneakers

IATA. (2023, June). *CORSIA Fact Sheet*. IATA.ORG. Retrieved May 29, 2023, from https://www.iata.org/en/iata-repository/pressroom/fact-sheets/fact-sheet---corsia/

IBIS World. (n.d.). Retrieved February 28, 2024, from https://www.ibisworld.com/industry-statistics/number-of-businesses/housing-developers-united-states

ICAO. (n.d.). *About ICAO*. ICAO. Retrieved May 30, 2023, from https://www.icao.int/about-icao/Pages/default.aspx

IEA. (2023, July 11). *Aviation - IEA*. International Energy Agency. Retrieved February 28, 2024, from https://www.iea.org/energy-system/transport/aviation

IMO. (n.d.). *About IMO*. International Maritime Organization (IMO). Retrieved May 30, 2023, from https://www.imo.org/

Kell, G. (2022, December 5). *From emissions cheater to climate leader: VW's journey from dieselgate to embracing e-mobility*. Forbes. Retrieved February 28, 2024, from https://www.forbes.com/sites/georgkell/2022/12/05/from-emissions-cheater-to-climate-leader-vws-journey-from-dieselgate-to-embracing-e-mobility/

Larsen, D. (2019, August 15). Opinion | The incontinent cows of middle-earth. *The New York Times*. Retrieved February 28, 2024, from https://www.nytimes.com/2019/08/15/opinion/new-zealand-cows.html

Mori Foundation. (n.d.). *Global power city index*. Retrieved February 28, 2024, from https://mori-m-foundation.or.jp/pdf/GPCI2023_summary.pdf

Neto, B., Dias, A., & Machado, M. (2012, October 23). Life cycle assessment of the supply chain of a Portuguese wine: From viticulture to distribution. *The International Journal of Life Cycle Assessment*, *18*. https://doi.org/10.1007/s11367-012-0518-4

New Zealand Wine. (n.d.a). *Sustainable Winegrowing New Zealand | New Zealand Wine*. New Zealand Winegrowers. Retrieved February 28, 2024, from https://www.nzwine.com/en/sustainability/swnz

New Zealand Wine. (n.d.b). *Sustainable Winegrowing NZ*. New Zealand Winegrowers. Retrieved May 19, 2023, from https://www.nzwine.com/en/sustainability/swnz/

NZ Wine Directory. (n.d.). *The wine regions of NZ*. New Zealand Wine Directory. Retrieved February 28, 2024, from https://nzwinedirectory.co.nz/wine-regions/

Pomranz, M. (2022, April 22). *New Zealand says nearly all of its wineries are now sustainable*. Food & Wine. Retrieved May 19, 2023, from https://www.foodandwine.com/news/new-zealand-wine-sustainability

The Port of Los Angeles. (2022, October 6). *Unprecedented year increases port emissions | Port of Los Angeles*. Port of Los Angeles. Retrieved February 28, 2024, from https://www.portoflosangeles.org/references/news_100622_aei

Rasmussen, N. (2023, May 3). *Liner supply falls as operators reduce average sailing speed by 4%*. Bimco. Retrieved February 28, 2024, from https://www.bimco.org/news-and-trends/market-reports/shipping-number-of-the-week/20230503-snow

Read, A. (2021, May 21). *Seaspiracy: Is the fishing industry killing our oceans?* Sustainable Food Trust. Retrieved February 28, 2024, from https://sustainablefoodtrust.org/news-views/seaspiracy-is-the-fishing-industry-killing-our-oceans/

Rupcic, L., Pierrat, E., Saavedra-Rubio, K., Thonemann, N., Ogugua, C., & Laurent, A. (2023). Environmental impacts in the civil aviation sector: Current state and guidance. *Transportation Research Part D: Transport and Environment*, *119*. https://doi.org/10.1016/j.trd.2023.103717

Savills. (2023, September 20). *Total value of global real estate: Property remains the world's biggest store of wealth*. Savills. Retrieved February 28, 2024, from https://www.savills.com/impacts/market-trends/the-total-value-of-global-real-estate-property-remains-the-worlds-biggest-store-of-wealth.html

Singapore Food Agency. (n.d.). *30 by 30*. Our Food Future. Retrieved February 28, 2024, from https://www.ourfoodfuture.gov.sg/30by30/

Statista. (n.d.). *Wine - New Zealand*. Statista. Retrieved February 28, 2024, from https://www.statista.com/outlook/cmo/alcoholic-drinks/wine/new-zealand#revenue

The Economist. (2010, March 23). *It's not easy seeming green*. The Economist. Retrieved February 28, 2024, from https://www.economist.com/international/2010/03/23/its-not-easy-seeming-green

Thiao, D., & Bunting, S. (2022). *Socio-economic and biological impacts of the fish-based feed industry for sub-Saharan Africa*. FAO. Retrieved February 28, 2024, from https://www.fao.org/3/cb7990en/cb7990en.pdf

Tibke, P. (n.d.). *Port and starboard [Updated 2024]*. Trade Finance Global. Retrieved February 28, 2024, from https://www.tradefinanceglobal.com/freight-forwarding/explained-port-and-starboard/

Times Colonist. (2020, May 3). *Trevor Hancock: Social tipping points, virtuous cascades - Victoria Times Colonist*. Times Colonist. Retrieved February 28, 2024, from https://www.timescolonist.com/islander/trevor-hancock-social-tipping-points-virtuous-cascades-4681046

UNEP. (2014). *The state of world fisheries and aquaculture*. UNEP. Retrieved May 30, 2023, from https://www.unep.org/resources/report/state-world-fisheries-and-aquaculture

Vakili, S., Ballini, F., Schönborn, A., Christodoulou, A., Dalakis, D., & Olcer, A. (2023, February 27). Assessing the macroeconomic and social impacts of slow steaming in shipping: a literature review on small island developing states and least developed countries. *Journal of Shipping and Trade*. https://doi.org/10.1186/s41072-023-00131-2

Vertical Oceans. (n.d.). *Vertical oceans*. Retrieved February 28, 2024, from https://shop.verticaloceans.blue/

WEF. (2022, December 9). *Can the aviation sector get to net-zero emissions by 2050?* The World Economic Forum. Retrieved May 29, 2023, from https://www.weforum.org/agenda/2022/12/aviation-net-zero-emissions

World Footwear. (2022, September 28). *All the facts and numbers of the footwear industry in 2021*. World Footwear. Retrieved February 28, 2024, from https://www.worldfootwear.com/news/all-the-facts-and-numbers-of-the-footwear-industry-in-2021/8152.html

Zhang, W. (2023, November 13). *China: number of real estate developers*. Statista. Retrieved February 28, 2024, from https://www.statista.com/statistics/242427/number-of-real-estate-developers-in-china/

CHAPTER 12

Reasons to Be Cheerful and Next Steps

"It is not the most intellectual of the species that survives; it is not the strongest that survives; but the species that survives is the one that is able best to adapt and adjust to the changing environment in which it finds itself." Charles Darwin

General Post Office, Ho Chi Minh City, Vietnam (authors own photograph)

S. Crispin, *Sustainability in the Built Environment*,
https://doi.org/10.1007/978-981-95-6549-8_12

The Good News

A vast effort from academics and across the value chain has already gone into working out how to decarbonise the buildings, construction and infrastructure that make up the built environment. This hard work has all been done and is freely available, this book has just put as much of this information as possible in one place. As Dhawal Doshi commented, "much of the work we have done remains invisible, everyone can see the transition to electric vehicles but no one sees a building chiller. As my wife says, it is similar to the invisible benefits of a yoga session compared to a gym workout". It's time we gave more recognition to the buildings that have taken up yoga.

Here are some of the key reasons to be cheerful:

1. Everything that we need to get the job done is already available.
2. The global community has the infrastructure, experience and frameworks to take action.
3. The cities in Africa show that renewable energy can be a significant contribution to a low carbon grid in developing countries.
4. The Singapore target is for 80% green buildings by 2030.
5. Pakistan embedding green building into local building codes.
6. China has implemented energy efficiency standards for buildings, introduced policy incentives and energy efficient building labels.
7. City level energy efficiency and greenhouse gas emissions standards for buildings in New York show that city level regulation can be effective.
8. Multiple developers that are taking real and measureable action to decarbonise embodied and operating carbon and sharing their efforts.
9. Rapid growth of green finance estimated CAGR of nearly 20% between 2025 and 2034.
10. The efforts already made to develop low carbon building materials with further work underway
11. Global market penetration of green building certifications.
12. Valuation guidance to explicitly reflect ESG actions.
13. Landlord and tenant collaboration to improve sustainability performance.
14. Development and adoption of technology to reduce carbon intensity of construction.

15. Wider knowledge and understanding of circular economy principles in building waste management.

Education and awareness has been a recurring topic in this book and that has been part of the rationale for this book. "It's clear there is an increasingly urgent responsibility for educators to provide education about sustainability. Whether it's business schools analysing the creation of non-exploitative supply chain practices, architecture schools encouraging net zero building designs or elementary schools introducing progressive consumer behaviour, all educators are in a position to contribute" according to Dr Paul Clark of Thompson Rivers University.

What remains little discussed are the systemic connections between all the reasons to be cheerful. They are interrelated. The connection between sustainable cities and public health, the connection between natural wetlands and flood prevention, the connection between façade audits and lower capacity chillers, the connection between indoor air quality and talent retention, the connection between energy efficiency and financial performance and many more.

The Cost of Doing Nothing

We can of course choose to do nothing and leave it to the politicians, after all, what can one individual do? The problem with politics in many countries, is the short-term electoral cycle inhibits action to deliver long-term sustainability goals. This is the tragedy of the commons scenario where self-interest leads to mutual destruction. Those of us who enjoy a comfortable modern life find it easy to think only of the impact on the wealthy city residents and forget the vast populations living in already marginal and sub-marginal environments. How many of those will stay put as droughts and floods worsen and famine follows? At some point, as past civilisation collapse shows, they will all be on the move in a desperate search for a new place to live.

The tragedy of the horizon means we can choose to not worry about it today. Just leave it to the next generation because "we can't afford it now". But we can't afford to do nothing either. The impact of climate change will not stop at the borders of the less developed countries. What happens as the floods, bushfires, subsidence, droughts increase in frequency and have a regular impact on the lives of those in developed countries too? How many uninsurable neighbourhoods does it take to make all homeowners take note and demand government action? How many elderly relatives are we willing to lose prematurely to heatwaves or extreme cold? We know this is already happening. Will we look back on this moment and wonder why we didn't do more, or will we shrug and say "how was I supposed to know? we left it to the politicians".

We may choose to "follow the data", but what we always do in time of crisis is listen to the scientists and their forecast scenarios. It is rational to act based on data-driven forecasts while knowing that forecasts aren't always 100% accurate. Fishing boats sensibly stay in harbour when a big storm is expected, and we take an umbrella in the morning because rain is forecast at 17:30. These are common actions we take based on a scientific forecast while still hoping for the best.

What Should Happen Now?

The truth is we all know what is happening to the environment, we can see ice caps and glaciers melting, almost all scientists agree on the reasons. Whether we choose to believe it is another matter. What we do not all know is the full extent to which the buildings we depend on for our economic and physical well-being are harming the environment and the health of future generations. What we need to do next is set out in brief below.

Acting together in small achievable steps, "the silver buckshot approach" of the Empire State Building, spreads the costs and shares the benefits. Reaching net zero across the entire built environment is the responsibility of all market participants, including residential homeowners.

The degree to which individuals and organisations can bring about change and the impact they can have varies considerably. We cannot expect a homeowner to be able to influence the cement used in the construction process any more than we can expect a glass maker to influence the electrification of a 40-year-old office building.

This makes things easier. Every individual participant just needs to focus on what they can do and not worry about the rest other than knowing that others are making their contribution too. There are significant interventions that can happen that will result in a cascade of impacts. For example, is there any reason why, over a period of time, all finance should not be green finance? After all, if the term of a loan or bond extends 2 or 3 decades into the uncertainty of our climate future climate risk should always be a factor. The cascade of collaborative sustainability action resulting from green finance would extend throughout the value chain.

Is there any reason why requirements for low carbon materials cannot be incorporated into building codes everywhere? Manufacturers would respond and implement the technologies they have been researching and testing, making use of green finance. Another cascade of action would follow.

Is there any reason why all new homes shouldn't come with solar panels installed and a storage battery? Then install them on existing homes over a period of time? A 5–10% reduction on grid energy in all developed countries would be a massive difference and give each home a certain amount of energy independence. Other changes such as insulation would have additional impacts and campaigns have been setting out these ideas with little success.

These are all questions for the policymakers and regulators, yet we cannot leave it to them alone. The industry needs to speak up more frequently and celebrate many more success stories far more loudly. The role of building industry and property owning lobbyists in the corridors of power should be questioned. At what point do the political risks from the cost of climate change exceed the benefits of political donations? Politicians who are complicit in doing nothing must be held to account. Compared to cars, aviation, plastics meat and others, the far more harmful building and construction industries are still largely lurking in the shadows. There needs to be an effective, equal and opposite lobby group that represents societal interests.

Some suggestions include:

1. GlobalABC to work towards universal recognition and support global and local policy makers, regulators and public and private industry stakeholders come together to understand, determine and adopt optimised sustainability practices for the built environment. At the same time be nimble enough adapt the approach for local markets.
2. Acceptance for action on the built environment needs to gain a level of understanding, acceptance and urgency similar to that seen at the time of universal global action on CFC's. For building cooling

systems, this should include a global data base of all existing refrigerants and implementation of life cycle refrigerant management systems.

3. All operational and management green certification systems to include publication of energy use intensity alongside the rating.
4. Policymakers and planners must adopt sustainability as a core requirement for planning and building approvals. To consider, at minimum, building orientation for optimised energy consumption (solar gain in colder seasons and shade in warmer seasons), building materials with recycled and recyclable content, structure and layouts for energy efficiency and operational energy efficiency plans. This includes:

    a. Regulations to cover operational efficiency considerations including mandatory energy consumption reporting and adoption of green lease partnership arrangements as a regulatory requirement.
    b. Publication of energy consumption of buildings to enable tenants to make informed decisions and for tenant related energy efficiency improvements to be identified.
    c. Policy, regulation and tax incentives to encourage deep retrofits of poor energy performing buildings.
    d. Maximise use of roof tops for solar power generation and requirements for energy providers to offer a feed in tariff at a rate based on wholesale producers, varied for volume.
    e. Mandatory green building certification of all commercial buildings and interior fit outs with emphasis on energy efficiency performance. An assessment like NABERS is suitable for wide adoption.
    f. Mandatory energy and façade audits and energy optimisation plans to be submitted to local government by building owners and tenants, to be updated every 3–5 years.

5. Fossil fuel free buildings to be adopted as standard among all government owned and occupied buildings.
6. Audit of trees and biodiversity in urban areas with preservation of "shade rights" for pedestrians alongside other improvements to protecting urban biodiversity.

7. Water conservation measures to include policy, regulation and tax incentive alignment to encourage grey water recycling, rainwater capture and water efficiency measures in all built environments.

Practical Matters

There are lots of positive actions that have already been implemented, some examples have been provided in preceding chapters. Simultaneously co-ordinating the scaling up of the innovative solutions while finalising the way around or over the last obstacles remains a particular challenge.

Resolve the policy gaps at city level by enshrining environmental requirements into updated building codes and with standards for embodied carbon and operating efficiency set into regulations. We know what the standards should be and some are doing this already, those cities that don't will lose competitiveness over time.

Positioning GlobalABC as the key platform for all built environment stakeholders from the building and construction sector to come together. GlobalABC is currently less well known than the IMO, many industry professionals have never heard of it. A Google search of GlobalABC yields around 77,000 results compared to 122 million for the International Maritime Organisation. The IMO has achieved international agreement to decarbonise shipping while GlobalABC is publishing reports asking why not enough is happening. Resources must be allocated to change this.

Establish a playbook of low cost and quick fix options to make brown buildings green on the inside, "Kiwi Fruit Buildings". This would include sustainability data management, building optimisation plans, lighting, façade audits, green leases, waste management, etc. A playbook adapted for local markets at city level as "guidance for building owners" should be integrated into city net zero pathways.

Take a systems approach to decarbonising the built environment. Mission Impossible Partnership looks at industrial silos, aluminium, concrete and cement, chemicals, steel, aviation, shipping and trucking. This is all very well, but they are mostly connected by a common buyer, a property developer, usually financed by a bank. Let's stop looking at siloed industries and start tackling the hard to abate industries as a green value chain by focusing on the key buy side; the developers of the buildings, their banks, their contractors who manage the procurement and all the various professional advisors.

All real estate debt should follow the green finance guidelines, if only for the lenders risk management. Investors and lenders should have the knowledge, understanding and confidence to identify and price the risk that goes with the different approaches to climate risk. Looking at the links between the built environment and climate change; property debt finance and financial crises it is not difficult to conclude that all real estate debt should be sustainable debt. Serious effort should be made to match an entire loan term with climate risks over the same period. If a mortgage repayment term is 25 years, the climate risk outlook should consider a 25-year forecast of scenarios. Likewise insurance premiums should consistently reflect specific climate risks and efforts to mitigate those risks. Insurers are already deeming some locations uninsurable. Those that doubt the level of climate risks can refer to insurance premium trends for an indication of the forecast losses rather than merely considering historic trends. Reluctant to act building owners should be paying close attention while those that are taking action should be recognised.

Achieving Industry-Wide Adoption

Simplification of what we measure, focus on the metrics that matter. Green standards, including ESG reporting itself, results in costly and complicated data sets across operations and building materials. The three that matter most are:

1. Embodied carbon.
2. Energy use intensity.
3. Renewable energy supplies.

We can do all three by taking a long-term view. Embodied carbon means material passports, energy use intensity means measuring and publishing data, renewable energy supplies means decarbonising the grid. Additional items to the above include measuring and managing total building carbon intensity, circularity for building materials, a data base of refrigerants in use and managing indoor and outdoor air quality.

A Final Key Question

A key last question remains: why is there so little discussion and knowledge about environmental sustainability and buildings? Compared to, say, vehicles, aviation, meat and so only, buildings barely gets mentioned in the media and public discourse. Visibility is certainly part of the answer, we see the buildings but not what goes into them. Dependency is another part, we are dependent on buildings for all our basic needs as well as our economic well-being.

We know the built environment accounts for 35–40% of GHG emissions and that the total global value of real estate makes it the world's biggest asset class. It is not difficult to imagine the collective weight of influence that goes with such wealth but is that really enough to hide the assets under a green cloak. To what extent is this wealth consciously or unconsciously keeping this subject in the dark and lobbying against regulations?

This book details what some of the most iconic structures and best known property developers and tenants are doing and they are far from alone. As discussed, impactful sustainability action is taking place in the built environment across the globe. At the same time, incumbent industry lobbying remains a challenge. This includes the energy sector as well as the building materials, construction and property development industries. Taken together, these are powerful lobby groups. In the absence of persistent political will, the key consistent message must focus on the business case as a means to overcome both the lobbyists and the political hurdles such as distribution of the transition costs.

Some of those that understand that sustainability results in increased values are implementing sustainable strategies. Those that don't or cannot are left behind potentially leaving their shareholders and debt holders with stranded assets to deal with. They may also be backing lobby groups in blocking industry-wide change. Lobbying may be conscious or unconscious (saying they're doing the right thing while actually doing something a bit different). This is a complex topic, simply finding common standards and definitions is far from easy. There are likely to be knowledge gaps about both valuation and sustainability and some of the relatively low cost and simple improvements that can be made. In some cultures, a reluctance to admit that current practices are not quite best practices may also be an issue.

All the difficult to tackle problems discussed in this book can be resolved, the evidence has been summarised in this book. Some solutions are already available and others are in progress, but scaling up and achieving universal adoption is the hurdle that remains. That is somewhat simplistic of course, but we can assume the current trajectory of innovation will continue. To tackle these last key questions, we need to talk more about the economic and environmental opportunity of the green building era as a single opportunity, "environomics" anyone? The case for both is clear and intertwined. Let's not allow the tragedy of the commons and the tragedy of the horizon become the tragedy of the buildings.

Appendix 1

Table A.1 Summary of GlobalABC "Why are We Still Not Adapting" Report

Section	*Title*	*Summary*	*Remarks*
1	We know the Impact of Climate Change	"Stakeholders within the building and construction sector possess the knowledge and tools necessary to tackle climate change head-on. They acknowledge the urgency of the issue, yet remain trapped in a frustrating cycle of slow progress in adaptive action"	In many markets, government show reluctance to take the lead through regulation. Building owners may know have the technical knowledge and tools but do they have the incentive? Do they understand the business case, including the risk of doing nothing?
1.A	The Impact is Global	Climate change is a global crisis with a wide range of impacts. It is an urgent priority requiring immediate mitigation and adaptation efforts	We know what is happening but the response is slow. Those that protest in support of the urgent actions called for by the science are frequently vilified in the media and sometimes jailed
1.B	The Buildings and Construction Sector is Particularly Impacted	The built environment supports both urban and rural economies and human life. Cities are built using carbon intensive resources and frequently result in heat islands. The built environment is also vulnerable to climate-related risks that affect human health and economics	Cities have driven economic prosperity. They are now at risk of being most exposed

(continued)

(continued)

Section	*Title*	*Summary*	*Remarks*
1.C	There is an Urgent Need for Resilience and Adaptation to Climate Change	Substantial adaptation efforts are required, many of which require government co-ordination. These include early warning systems, flood defences, government policies, etc. Active participation from communities, public and private sectors, NGOs, knowledge management and communication are required	The knowledge we need exists but the understanding, communications and willingness to act seem to be missing. Governments seem too content to sit back and take a light touch approach
1.D	The Return on Investment on Adaptation Measures is Real and Significant	The cost of both adaptation and mitigation are moderate compared to the annual total cost reduction of US$100 billion and a reduction of US$183 billion in annual unavoidable damages. The sooner mitigation is done the bigger the potential savings	The costs of adaptation fall on individual asset owners but the savings may be uncertain or too distant. The benefits of mitigation may be perceived as being transferred to others regardless of whether they make investments or not. The concepts are unfamiliar, and the benefits are uncertain while the costs are real and current
1.E	The Pace of Adaptation is Still Insufficient	The funding gap for adaptation in developing countries is estimated to be 5 to 10 times current funding and the gap is growing. Private finance and investors should place greater emphasis on adaptation and resilience compliant assets through "adaptation-aligned finance"	Asset valuations, insurance and finance costs need to fully reflect long-term climate-related financial risks related to inaction. Investment and other professionals need to fully understand the issues and integrate into financial modelling and acquisition due diligence
2	Stakeholders Already Have the Tools to Successfully Address Climate Adaptation	Understanding stakeholders across the value chain is the key to specific changes that each can make. "10 Principles for Effective Action" enables a multi-disciplinary cross sector stakeholder approach. Different stakeholder groups have different skill sets and ability to address specific risks	Collaborative action is the key to implementing changes. Stakeholders are used to their silos or focus areas and reluctant to reach across. Professionals are trained to do what they are good at; stakeholder training does not encourage multi-disciplinary understanding. The inability to prove what financial benefits accrue by mitigating tomorrow's risks today is a key gap

(continued)

(continued)

Section	Title	Summary	Remarks
2.A	National Governments	National governments cannot make decisions for business and nor can they impose the financial cost of adapting buildings on taxpayers. Their role is in setting a policy framework that supports adaptation and mitigation, provision of information, resources and guidance, supporting research and directing public and private investment. Inclusive resilience preparedness is part of their key responsibility to protect the safety and well-being of citizens	All governments require a mandate. A common perception that decarbonisation means loss of amenity or higher costs makes the role of government in setting policy and implementing regulations at all levels very difficult. Transparency and consistent communication are requirements that challenge all elected and appointed government personnel. Simple steps include sustainability data transparency, building code updates, certification standards. Sustainable built environments requires collaboration and working across departments in ways that many sometimes find challenging
2.B	Local Authorities	Well-being and responses to extreme weather are local issues. Adaptation to improve resilience reduces rebuilding and recovery costs. Regulations that National governments find difficult to introduce are often more readily deployed by local authorities. Local risks need contextual responses, assuming jurisdiction for collective responses and resource allocation are more straightforward at the local level. At the same time, they face financial and governance constraints	The combination of the burden of first response to disaster relief and the lack of national policy and regulations means local Authorities and State/Province or City level are in the front line. The local nature of climate-related disasters and the geographic and demographic variables means that contextual regulations for adaptation of resilience are effective at the local level
2.C	Investors and Asset Managers	Real estate assets are exposed to multiple extreme weather risks. Damage has to be paid for and the threat of value loss is real. Operating and insurance costs may increase without adaptive strategies to minimise risk exposure. Investors and asset managers have to learn about the risks and adaptation strategies. Taking a proactive approach to future proof assets is already a requirement to avoid escalation of economic costs related to asset repricing	Climate risk is financial risk and buildings are generally considered a stable, long-term income producing asset class. With risk comes opportunity and investors are already positioning themselves and climate risk specialists The need to avoided asset stranding applies not only for the asset itself but also for the asset manager and their future income. It is worth bearing in mind the frequency of real estate-related financial crises

(continued)

(continued)

Section	*Title*	*Summary*	*Remarks*
2.D	Property Developers	Investors and tenants increasingly seek climate resilient buildings, regardless of asset class. Lower financial risk and insurance premiums are an obvious short-term financial benefit. Lower cost of repairs and transition costs are other financial considerations Property developers face the risk of changing regulations during the period between site acquisition and building work commencement. Regulations may also require expensive upgrades during the period of ownership. Climate resilience is "a strategic necessity" A highlighted risk is the impact of drought on the soil structure. This can result in subsidence and be costly to repair. These risks need to be identified at post development transaction due diligence Another risk relates to energy efficient buildings and the risk of damp walls and mould or overheating if indoor climate balances are not maintained Property developers also act has policy influencers. They are able to provide valuable feedback as a key stakeholder group	Arguably energy efficiency is a more important consideration than climate resilience. Any office tenant with a net zero target is very much focused on maximising energy efficiency as a crucial first step. With an inability to secure green tariffs, reducing electricity consumption is even more important. Investors may consider a wider range of sustainability factors but energy efficiency is likely to be high among them The risk of regulatory changes during construction or ownership should also be considered alongside the role of property developers as key stakeholders to be considered in policy development. This may vary widely in different governance jurisdictions and level of government. The role of property developers in political patronage is a counter factor. Under some governance structures, property developers may perfectly legally fund political parties or individual politicians. They are frequently significant taxpayers, employers and sponsors of culture and the arts. As land buyers then also fund governments. Any and all of these considerations ensures their seat at the top table Among tenants, there are variables such as asset class, location specific risks, tenant size and multi-tenant buildings vs single tenant buildings On the matter of soil shrinkage in drought conditions, raft foundations are a common quick and cheap structural technique in locations where bedrock is limited, such as Shanghai. Foundation settlement have been shown to increase when ground water levels decrease (Amomfa et al., 2023)

(continued)

(continued)

Section	*Title*	*Summary*	*Remarks*
2.E	Insurers and Reinsurers	Promoting climate resilient buildings helps insurance and reinsurers reduce potential payouts. Relying on historical data is of limited use in assessing future climate risk. Underestimating future climate-related risks can result in losses that put insurer solvency at risk. Increasing severity and frequency of events is particularly hard to predict. Insurance policy documents and working with policymakers are the principle means of mitigating risk exposure	Pricing of future climate risk is impossible to do with any degree of accuracy. Insurers can be more assertive in insisting their clients take action to adapt buildings or penalise them through pricing mechanisms. A green discount and a brown premium is the reverse of the revenue equivalents. Relating the cost of insurance to clear decarbonisation progress is not directly linked to mitigating climate change but when implemented across all policy holders it can be an effective mechanism to transform markets
2.F	Architects	Architects can help move their clients away from meeting minimum sustainable design incorporated into building codes and advocate for more climate resilient buildings. Incorporating energy efficient design, natural materials, building orientation, shading, passive ventilation and so on can make buildings more comfortable and energy efficient. Adaptive features can be incorporated into existing structures. Resilience to heatwaves, flooding, typhoons are already basic design features	Put an end to "design as usual" and implement "design as unusual" practices. Innovative designs are required in both new buildings and to retrofit existing buildings Incorporating resilient design needs collaboration with tenants or multiple owners and can become extremely complex with multiple owners of a common asset in the case of a strata title owned building Including a wider range of professional advisors in 2.F and 2.G, as described in table xx below would allow for a greater range of influences to be incorporated into this stakeholder group Quantity surveyors to work on life cycle costs and cost implications of sustainable choices. Include maintenance and durability considerations Valuers to incorporate price and rent premiums for green buildings Finance professionals and accountants to recognise the value added in making loans and in audit valuation reports

(continued)

(continued)

Section	*Title*	*Summary*	*Remarks*
2.G	Engineers	Building codes and regulations must be updated to consider the needs of the coming years Must use adaptive designs and improve maintenance planning to ensure safety, sustainability and resilience of the built environment Various sustainable design strategies include nature based designs to include natural elements, modular design for easier upgrading or replacement, transformative and redundant design to allow for change of function in case or change in function or failure in extremes Engineers also need to be ready to address maintenance planning and increased maintenance costs. Changing climates can be expected to result in higher provision for maintenance costs for infrastructure and buildings	Building engineers to design and specify the plant and machinery and ensure it is optimised at time of building commissioning and regularly maintained Civil engineers should be given greater role in adaptation and the means to achieve it Working together with insurers can help prepare resilience planning and recovery. This may extend to the wider community as well as the building occupants
2.H	Material, Equipment and Construction	There is an opportunity for this stakeholder group to supply sustainable materials, equipment and construction methods. In particular staying ahead of regulations will help suppliers to keep a competitive edge and enhance reputation in the market There may be conflicting needs for new materials with complex processes and the need to be sustainable (recycled content and recyclable, low carbon footprint, etc.) The market needs to understand the new materials and adapt to their use	Scaling up production of sustainable materials and equipment can make use of green loans and blended finance to help achieve economies of scale Understanding the trade-offs in sustainable materials, equipment and construction can be a difficult issue with complex calculations and a lack of clear answers Material performance in the long-term vs short-term cost, maintenance issues, aesthetics, costs and so on now need to include environmental impact. Decision-makers too need to understand the choices and trade-offs

(continued)

(continued)

Section	*Title*	*Summary*	*Remarks*
2.I	Property and Facility Managers (PFM)	Climate change impacts property values, repair and replacement costs, operating costs, business continuity and occupant safety and well-being Regulatory compliance is a key role and can help reduce costs and insurance premiums when proactively managed Key roles include: 1. Understanding the building operations and assessing vulnerabilities to inform owners 2. Collecting data to inform decision-making on adaptations and upgrades 3. Working with a range of stakeholders to inform and advise 4. Work with suppliers to prioritise sustainable choices in design and operations The PFM businesses can also enhance their brand value by adopting sustainable practices	The various roles in PFM include cleaners, landscaping, pest control, waste management, security, maintenance, finance, leasing and more. Taken together with the design led engineering teams they play a crucial role in protecting the property, business continuity, post extreme weather clean ups and operations Individually these tasks may appear simple but delivering them together efficiently is a complex operation that is often overlooked. It now has the additional burden of being in the front line of climate change adaptation and extreme weather response management
2.J	Occupants, Private Owners and Civil Society	Occupants include owners and tenants. Their behaviour can save energy, reduce waste and use sustainable materials. They have the right to expect a clean, healthy and sustainable environment. Resilient and sustainable buildings can expect higher values and reduced operating and maintenance costs Civil society channels the drive for changes in the built environment to not only adapt buildings but to improve substandard housing that is less resilient in the face of extreme weather. Civil society also plays a role in monitoring law making, transparency, accountability and oversight Trusted sources of information act as messengers to disseminate knowledge. This might include contractors, insurance agents, planning authorities and even neighbours to support collaboration to promote sustainability and resilience	This is a diverse stakeholder group. In the commercial property sector the business tenant tends to be the key group. Commercial building owners are a diverse group that includes smaller and localised property developers to major listed developers and international funds The residential sector includes private individual, property funds, local authorities and private and listed property developers among others Civil society has to account for a wide range of interests but the main obligation would seem to be marginalised groups who are at the highest risk and to expect law makers to enact building codes and regulations that require investment in adaptation

(continued)

(continued)

Section	*Title*	*Summary*	*Remarks*
3		Why Are We Not Seeing More Widespread Adaptation	
3.A	Lack of Incentive to Bear the Initial Cost	The complexity of the incentives, short-term return and economic considerations mean that the stakeholders across the value chain are reluctant to unilaterally embrace change The imperative of the financial return for the owners means professional advisors are reluctant to advocate for increased costs associated with sustainability or resilience despite the longer term cost benefits Buildings that stand for 80 years, or even centuries, will typically have multiple owners and a range of uses Reducing costs at the initial stage gives the first owner security of profit after the high risk development stage is complete. It does transfer the cost of upgrade to subsequent owners and users at the end of the value chain	Turning away from systemic incentive dysfunctionality through a cascade of interventions to a sustainable model for the built environment is the ultimate barrier to overcome Unlocking the barriers to improve normalisation of sustainable and resilient design would result in mainstreaming of action across the built environment. At the same time scaling up use of sustainable solutions would result in reduced costs Green finance is another possibility to support the additional investment required for green building technologies The biggest challenge remains the stock of existing buildings and needs a different approach to that for ground up new development. There are also different approaches for different building types and ownership structures

(continued)

(continued)

Section	*Title*	*Summary*	*Remarks*
3.B	Design Life vs Service Life	The longer a building lasts, the lower the environmental impact. The most common designed life of a building is 50 years and this is what is used in the life cycle assessment. The main factors being the structural integrity of the building as determined by the concrete, steel reinforcement and corrosion as well as maintenance If designed life is extended from 50 to 80 years the environmental impact is reduced by 29%. Or by 38% and 44% for lifespans of 100 and 120 years respectively Premature demolition occurs when buildings are replaced before the end of their physical life usually due to "subjective perceptions" or changes in use	In practice many modern buildings last more than 50 years anyway. The 102-storey Empire State Building was completed in 1931 so not far off double what might the design life of a modern equivalent Ironbridge in the town named after it in Shropshire England dates from 1779 and was the first fully iron bridge In Spain catastrophic floods killed over 200 people after torrential rain in November 2024. The scale of the disaster surpassed the death tolls in Germany and Belgium in 2021 (Jopson, 2024). One silver lining was a Roman Dam, dating from the first-century AD, which was credited with diverting the waters from the town of Almonacid de la Cuba (In Spain News, 2024) Oriel Chambers, Liverpool, England dates from 1864 and features the first metal-framed, glass curtain wall (Oriel Chambers, n.d.). As a Grade I listed building it will be maintained to stand indefinitely Modern built environments are enriched by structures that have extended their designed life time. One of the consequences of rapid urbanisation will unfortunately be premature demolition of poorly designed and cheaply built buildings

(continued)

(continued)

Section	*Title*	*Summary*	*Remarks*
3.C	Historical Bias and Hazard Data in Codes and Standards	Older buildings were not designed for current climate conditions or for future climates. New buildings must integrate predicted climates into their design, even anticipating changes to building codes and standards	Future proofing is not a new concept to the building and construction industry, frequently alternative use will be considered. Older office buildings may be converted for residential use for example Applying this to climate resilience is newer but not entirely. Storm prone cities frequently have stepped entrances to subway stations and ground floor doorways. Various additional elements of sustainable design include building orientation, shading, green walls and so on
3.D	Lack of Measurable, Stepwise Action (As Opposed to Mitigation)	Adaptation refers to preparing and responding to a changing climate, mitigation involves reducing emissions and the causes of climate change Adaptation requires investment that meets a standard for resilience. That requires commitment in response to a clear understanding of the need. Mitigation can be implemented over a period of time, such as changing to LED lightbulbs as the old ones fail	The return on investment for adaptation for indeterminate risks is difficult to quantify with certainty. The assumptions of future flood event frequency may fall within a margin of error but there is no certainty that will ever happen or the extent of the damage may be overstated. There are unknowns Mitigation is easier to apply to a financial model. Investing $x in energy reduction results in a saving of $y. The decision is easy
3.E	Competing Interests	Losses from environmental disasters tend to be in urban areas. The impact may be felt hardest in informal or low-cost settlements. Enhancing resilience to mitigate the impacts will result in increasing property values meaning the housing is no longer affordable and the populations such upgrades are intended to help have to move away. Housing affordability and resilience are in conflict. Balancing competing interests is the job of policy makers through collaboration and innovation	The conflicts inherent in sustainability and climate change leave any action open to criticism. Sometimes it seems everyone knows better than the scientists who have dedicated their lives to understanding the topics Collaborative frameworks do exist but in the built environment we don't see them being used to the same extent as, say the finance industry or shipping even

(continued)

(continued)

Section	Title	Summary	Remarks
3.F	A Primarily Local Issue	Adaptation requirements depend on the local context requiring a bottom-up approach. Mitigation is better suited to standardisation and a top-down approach. Local issues come with a complex set of barriers based on Piggott-McKella and McNamara's classification: 1. Cognitive and individual barriers based on attitudes, beliefs and knowledge. There is a lack of reliable information 2. Regulatory and institutional barriers include financial constraints and regulations 3. Socio-cultural barriers might include informal housing, existing infrastructure, past planning, the more immediate needs of less well-off communities to support their daily living needs (Piggott-McKella et al., 2019)	The barrier of local issues means multiple locations have to take action, not a single national authority. This means waiting for each to reach a level of awareness, knowledge and understanding to implement their own local responses and overcome the various hurdles described The knowledge gap can be resolved through raising awareness and publication of reliable information. The information required is generally available Financial and regulatory hurdles are being overcome but it takes time for responses to be developed, planned and implemented The socio-cultural barriers require more sensitive handling. It is not always simply a case of NIMBYism, people's lives and livelihoods are at stake

(Pianet, 2024)

Appendix 2

Table A.2 Selected built environment professional associations and their public sustainability comments

Professional Associations	*Sector*	*Description*	*Sustainability agenda*
AIA: The American Institute of Architects	Architecture and design	· "The largest, most influential network of architects and design professionals"	· AIA 2030 Commitment provides standards for net zero emissions in the built environment

(continued)

(continued)

Professional Associations	*Sector*	*Description*	*Sustainability agenda*
ANREV: Asian Association for Investors in Non-Listed Real Estate Vehicles	Non-listed real estate funds	· Forum for investors their managers and service providers to help development of the sector as a competitive sustainable asset class · Global agenda of transparency, accessibility, professionalism and harmonisation across the non-listed real estate industry	· Recognises the role the industry has on the transition to a more sustainable future · Supports and promotes ESG reporting aligned with best practices · Encourages members to integrate ESG into their business practice at asset, portfolio and organisation level.
	Securitised real asset investors	· Community of professional leaders from top publicly-held and privately-owned real assets ownership, development and management firms	· APREA champions the adoption of sustainability and ESG best practices among members · Aims to raise awareness and stimulate action towards a low carbon built environment · ESG KnowHow offers sustainability-focused research, tools and resources
BOMA: Building Owners and Managers Association International	Property developers, landlords and property managers	· "Represents the owners, managers, service providers and other property professionals of all commercial building types"	· BOMA International has been promoting sustainable buildings and energy efficiency for more than a decade. This includes advocacy, scholarships, building codes, and energy performance standards and awards

(continued)

(continued)

Professional Associations	*Sector*	*Description*	*Sustainability agenda*
CABE: Chartered Association of Building Engineers	Professional engineers who specialise in the built environment	· A professional body of members who work across the built environment life cycle to create better, safer and more sustainable buildings by sharing knowledge, developing professionals and raising standards	· Provides knowledge sharing and on line training about sustainability related topics
CORENET	Corporate real estate professionals association	· Represents around 10,000 members with responsibility for real estate assets of large corporations	· Provides online information and training resources. Corporate Real Estate 2030 is a research project including ESG as 1 of the 5 primary domains
IEMA: Institute of Environmental Management and Assessment	Environment and sustainability	· Professional body for those working, studying or interested in sustainability across all industries	· ISO14001 certified, committed to Pledge to Net Zero and progressive reduction of residual GHG emissions
RIBA: Royal Institute of British Architects	Architecture	· Global membership body for architects to promote better places and buildings, stronger communities and a sustainable environment	· A signatory to the UN Global Compact promoting best practice in the environment, human and labour rights, and anti-corruption

(continued)

(continued)

Professional Associations	*Sector*	*Description*	*Sustainability agenda*
RICS: Royal Institution of Chartered Surveyors	Property and built environment	· Advance knowledge, uphold standards and inspire current and future professionals. Create and protect natural and built environments · Set qualifications and standards for professionals working in the industry · Guide decision-maker and governments	· Goal to lead and influence sustainability · Focus to decarbonise the built environment, embed circularity, protect biodiversity and build climate resilience · Participant in 2024 inaugural Buildings and Climate Forum
ULI: Urban Land Institute	Land use and real estate	· Global network of cross-disciplinary real estate and land use specialists · Shape the future of the industry and create thriving communities · Set standards of excellence in development practice	· Global mission to decarbonise real estate · Goal for net zero operations by 2050 with meaningful decarbonisation by 2030 · Range of resources for members to understand the business case for net zero and implementation tools

Source (AIA, 2024; ANREV, n.d.; APREA, n.d.; BOMA, n.d.; CABE, n.d.; Corenet, n.d.; IEMA, n.d.; RIBA, n.d.; RICS, 2024; ULI, n.d.)

References

AIA. (2024, September 10). The American Institute of Architects: AIA. Retrieved November 14, 2024, from https://www.aia.org/

ANREV. (n.d.). *About ANREV*. ANREV. Retrieved November 14, 2024, from https://www.anrev.org/en/about-anrev/overview/

Amomfa, K., Quang, H. T., & Tuan, T. V. (2023, March). Effect of groundwater level change on piled raft foundation in Hong Chi Minh City, Vietnam using 3D-FEM. *Geomechanics and Engineering*, *32*(4), 387–396. https://doi.org/10.12989/gae.2023.32.4.387

APREA. (n.d.). *About*. APREA Asia. Retrieved November 14, 2024, from https://www.aprea.asia/index.php

BOMA. (n.d.). *BOMA Home*. Retrieved November 14, 2024, from https://www.boma.org/

CABE. (n.d.). Chartered Association of Building Engineers. Retrieved November 14, 2024, from https://cbuilde.com/

Corenet. (n.d.). Homepage. Retrieved November 14, 2024, from https://www.corenetglobal.org/

IEMA. (n.d.). IEMA - Institute of Environmental Management and Assessment. Retrieved November 14, 2024, from https://www.iema.net/

Jopson, B. (2024, November 2). *Spain deploys 5,000 more troops to boost flood rescue effort*. Financial Times. Retrieved November 14, 2024, from https://www.ft.com/content/5cb141d7-9c11-44bd-a8d1-c8d731662890

Pianet, V. (2024, March). *Achieving resilience in the built environment, why are we not adapting?* GlobalABC. Retrieved March 5, 2024, from https://globalabc.org/sites/default/files/2024-03/GlobalABC-White%20Paper-Why%20are%20we%20not%20adapting-FULL.pdf

RIBA. (n.d.). *RIBA - About the Royal Institute of British Architects*. Royal Institute of British Architects. Retrieved November 14, 2024, from https://www.architecture.com/about

RICS. (2024). *About RICS*. RICS. Retrieved May 29, 2024, from https://www.rics.org/about-rics

ULI. (n.d.). *About ULI*. Urban Land Institute. Retrieved November 14, 2024, from https://uli.org/about-uli/

The manufacturer's authorised representative in the EU is Springer Nature Customer Service Centre GmbH, Europaplatz 3, 69115 Heidelberg, Germany. If you have any concerns regarding our products, please contact ProductSafety@springernature.com

Printed and bound by CPI Group (UK) Ltd, Croydon, CR0 4YY

21/09/2026

02219945-0010